DESIGN AND ANALYSIS OF INTEGRATED MANUFACTURING SYSTEMS

W. Dale Compton, *Editor*

NATIONAL ACADEMY OF ENGINEERING

NATIONAL ACADEMY PRESS
Washington, D.C.   1988

NATIONAL ACADEMY PRESS   2101 CONSTITUTION AVENUE, NW   WASHINGTON, DC 20418

The National Academy of Engineering was established in 1964, under the charter of the National Academy of Sciences, as a parallel organization of outstanding engineers. It is autonomous in its administration and in the selection of its members, sharing with the National Academy of Sciences the responsibility for advising the federal government. The National Academy of Engineering also sponsors engineering programs aimed at meeting national needs, encourages education and research, and recognizes the superior achievements of engineers. Dr. Robert M. White is president of the National Academy of Engineering.

This publication has been reviewed by a group other than the authors according to procedures approved by a Report Review Committee. The interpretations and conclusions in this publication are those of the authors and do not purport to represent the views of the council, officers, or staff of the National Academy of Engineering.

Funds for the National Academy of Engineering's Conference "Design and Analysis of Integrated Manufacturing Systems: Status, Issues, and Opportunities" were provided by the Academy's Technology Agenda Program.

Library of Congress Cataloging-in-Publication Data

Design and analysis of integrated manufacturing systems.

  Papers presented, in part, at a conference held Feb. 25–27, 1987.
  Includes index.
  1. Computer integrated manufacturing systems—Congresses.   2. Flexible manufacturing systems—Congresses.   I. Compton, W. Dale.   II. National Academy of Engineering.
TS155.6.D47   1988   670.42′7   88-1766
ISBN 0-309-03844-8

Printed in the United States of America

CONTENTS

PREFACE

The manufacturing sector has been a strong and consistent contributor to the economic vitality of the United States. With the onslaught of imports from foreign manufacturers who have a cost advantage in either labor or natural resources, it is increasingly difficult for U.S.-based manufacturers to maintain their market share and sustain current levels of employment. It is apparent that the advantages of the overseas suppliers will be overcome only through a concerted effort that improves the efficiency of U.S. operations while providing high-quality products at a comparable price. This need to increase productivity, improve quality, and reduce costs has led many companies to search for new ways of achieving the conversion of a product idea into a manufactured product.

Recognizing that all participants in the enterprise—the product designer, the material handler, the material processor, the assembler, the service personnel, and the corporate management—contribute to the success of the operation, the search for improved ways to create competitive products must affect them all. We can improve the competitiveness of U.S. manufacturing only when we have come to understand the factors that affect the productivity of each of the segments of the manufacturing enterprise as well as the interaction among them. This search for new understanding implies a special need to improve the tools that are used to analyze and design systems of the complexity of a manufacturing enterprise.

It was recognition of the challenges in meeting this need that led the National

Academy of Engineering to hold a conference entitled "Design and Analysis of Integrated Manufacturing Systems: Status, Issues, and Opportunities" on February 25–27, 1987. The papers contained in this volume were presented, in part, at that conference. In this volume the authors, as experts in key areas of the field, explore the status of the tools that are used to design and analyze integrated manufacturing systems, and they identify many issues that arise in the use of these tools in designing and later in controlling these systems. Finally, each author has identified a series of research opportunities that should be explored as we search for better and more powerful tools for handling the problems posed by systems of this complexity.

Although each of the conference presentations generated intense discussion, the central theme of discussions throughout the conference was that U.S. firms and researchers interested in manufacturing must abandon the narrow view of manufacturing—the view from the perspective of a single element of the enterprise. They must learn to treat the entire system as an interacting set of elements that cannot be optimized in a narrow context. There is a corresponding need to abandon compartmentalization in manufacturing and to break down the barriers between design and its realization. Old patterns of limited interaction between elements of the manufacturing enterprise must give way to new patterns emphasizing communication and teamwork. We can no longer afford the disruptions and the inefficiencies that result when one unit throws a design, analysis, or test "over-the-wall" to another unit.

In addition to the topics that are presented in the papers included in this volume, it should be noted that discussions in small workshops led to a consensus that a more intensive search should be undertaken for the elements of a manufacturing system discipline. It was felt that both the practitioner and the educator would benefit from this—the practitioner from the generic tools that would assist him in analyzing, designing, and controlling systems and the educator from the availability of a systematic body of knowledge for presentation in the classroom.

On behalf of the National Academy of Engineering, I would like to thank especially the conference advisory committee (listed on page 225) that designed the conference, and the speakers at the conference and the individuals who attended and participated so actively in the discussions. With regard to the preparation of the manuscript for publication, I am especially grateful to H. Dale Langford, the National Academy of Engineering's editor, and Mary J. Ball, administrative secretary in the National Academy of Engineering Program Office.

W. DALE COMPTON
Senior Fellow
National Academy of Engineering

DESIGN AND ANALYSIS
OF INTEGRATED
MANUFACTURING
SYSTEMS

INTEGRATED MANUFACTURING SYSTEMS: AN OVERVIEW

JAMES J. SOLBERG

Manufacturing stands on the threshold of a new era in which all manufacturing enterprises must compete in a global economy. A gloomy view of this new economic environment can project dismal consequences for the quality of life in America. A more optimistic view recognizes that despite challenges to the nation's economic and technological strength, this fundamental change in the competitive environment presents opportunities as well as challenges. Indeed, it is possible to imagine a future of economic growth and prosperity on a global scale.

One thing that can be said with certainty is that all sectors of society will undergo profound changes. Consequently, many manufacturing practices that were effective in the past will no longer be so in the future. Those companies that prosper and those that fail may well be distinguished principally by their ability to plan for

change. Moreover, we can expect the economic environment to remain highly dynamic, so changes must be confronted with the expectation that they will be continual.

Fortunately, there are many options open to companies that understand the dynamics of this situation. The problem that they face is to identify, from among the rich universe of possibilities, those opportunities that represent the best investment of limited resources.

There are many conflicting voices speaking to manufacturers. Some emphasize automation equipment; others say that the data network is key. Noting delays in process flows, some suggest that attention to material handling is essential; others argue that it is only the processing steps that add value. Some put the primary burden for reform on management, some on the workers, some on suppliers, and so forth. CAD, CAM, CAE, JIT, SQC, MAP, MRP—the

1

list of abbreviations for promised solutions—goes on and on. How can we make sense out of all of this?

Numerous conferences and groups have addressed issues related to the changing nature of manufacturing. The National Research Council, through the Manufacturing Studies Board, recently completed a broad examination of the current status of U.S. manufacturing, covering technology, employment, and government policy (NRC, 1986). Several workshops have attempted to define research needs in more specific technical areas (ASME, 1986; NRC, 1984; Thomas, 1983; Volz and Naylor, 1985). Each of these reports provides useful and important background material.

The papers contained in this volume were prepared for a conference that was convened by the National Academy of Engineering to assess the strengths and weaknesses of current information and tools for planning and controlling integrated manufacturing systems, and to identify critical needs and opportunities for future study. Although the scope of inquiry included all discrete product manufacturing, it was limited to the technical aspects of the problem.

In the conference the emphasis on tools for planning and controlling operations reflected a concern for the adequacy of available methodology. The attention given to integration indicated that existing manufacturing technology is too segmented generally. With fragmentation of the research community mirroring the situation in industry, a principal objective of the conference was the creation of a synthesis of needs and understanding from a wide range of individual views. The dialogue between representatives of the user community who are involved in manufacturing and members of the research community in universities, industry, and other laboratories generated a deeper understanding of needs and opportunities.

The organization of the papers in this volume reflects the structure of the confer-

ence. In their paper, Erich Bloch, director of the National Science Foundation, and Kathy Prager Conrad emphasize the importance of gaining an understanding of current needs and future prospects in integrated manufacturing. The next two papers are authored by representatives from companies who had achieved notable success with automation. Arthur J. Roch, Jr., LTV Aircraft Products Group, describes the LTV experience with flexible machining and Laurence C. Seifert, AT&T Network Systems, discusses the AT&T integrated manufacturing system for electronics. The systems aspects of integrated manufacturing systems are described by Ulrich Flatau, Digital Equipment Corporation, and by John A. White, Georgia Institute of Technology. The human role in these systems is discussed by William B. Rouse, Search Technology, Inc. Technical issues in product design are discussed by Herbert B. Voelcker, Cornell University, and Daniel E. Whitney and coworkers, Charles Stark Draper Laboratory. The issues that relate to processing are discussed by Vijay A. Tipnis, Georgia Institute of Technology. System design issues are discussed by Rajan Suri, University of Wisconsin, and by Arch W. Naylor and Richard A. Volz, University of Michigan. System operation is discussed by Floyd H. Grant, FACTROL, Inc.

Not surprisingly, the papers present differing points of view. Also as expected, the authors tend to propose their own perspectives as the correct ones from which to address all other issues. Both the speakers and the participants recognized this during the conference and openly acknowledged their biases with good-natured humor. More than once, participants likened themselves to the blind men trying to describe an elephant based on which part of the animal each of them explored.

Although a full consensus does not emerge from this set of papers, important issues that transcend the narrow view of any individual discipline are clearly identified. The recurrence of these themes conveys a sense of unity on key issues. The first of these issues is the communication gap between industry and universities. Both groups acknowledge past neglect and express an eagerness to remedy the situation.

Industrial and academic representatives have identified the design-to-manufacturing transition as a second key problem. "Collaborative manufacturing" and "simultaneous engineering" are recently popularized phrases for a very old need, which nevertheless remains as one of the major challenges. Competitive pressures have increased the urgency of the plea for better coordination between design and manufacturing and for improved tools to support this coordination. To date, the research community has provided little help in this arena, leaving the practitioners to struggle as best they can with ad hoc procedures. The creation of a helpful methodology will be difficult, but at least the problem is now more universally acknowledged.

Also apparent from the papers is the extreme complexity of the tangled web of relationships among the various subsystems involved in manufacturing. The differing perspectives presented in the papers leave little doubt that, although important tools and insights are included by each author, none addresses all of the important relationships. One of the most interesting points of discussion in the conference was the question of how best to address this issue of complexity. Modern manufacturing systems are highly complex. They share some of the complexities of biological or economic systems, such as the interactions of an enormous number of variables whose relationships are unknown and a mixture of time scales, including very short to very long. The physical, logical, and temporal separation of consequences of actions often confounds direct attempts to control behavior. Unexpected side effects may counter the very effect we desire to achieve. One approach to this problem is to acknowledge

that such complexity is inevitable and, therefore, to concentrate on developing far more sophisticated methods of analysis and control. An alternative approach is to conclude that we will never learn to understand and manage all details of such complicated systems, so we must strive to simplify them.

To a neutral observer, it is obvious that both approaches should be pursued as vigorously as possible. In fact, this debate over whether it is better to master complexity or to avoid it is somewhat reminiscent of the perpetual debate over hardware and software issues. Both approaches are valid, and neither can advance for long without progress from the other. Still, we can expect the debate to continue for years.

The general tone of the papers contained in this volume reflects a sense of the change that is taking place. The old vocabulary of manufacturing systems seems inadequate for expressing many of the new concepts in the field. The birth of new paradigms of thought is always a bit unsettling, so it would not have been surprising if the broad range of papers included in this volume included divergent views. That this is not so indicates general agreement that serious changes in the direction of manufacturing are in order.

Under these circumstances, it is important that we understand both the global issues associated with manufacturing as well as specific technical needs and opportunities. Two questions might be asked. Why is manufacturing different now? Of the knowledge and methods we have acquired in the past, what should be kept and what should be abandoned? In searching for answers to these questions, we must not be too willing to discard what has been done in the past, for there is a danger of chasing a passing fad or misreading a cyclical change as permanent. At the same time, we cannot hope to make significant progress in the competitive race if we insist on clinging too firmly to traditional thinking.

Such issues as product cost and quality are, of course, timeless. The ability to produce what the customer wants in a timely manner is also basic. These fundamental driving forces point to a continued need to improve the efficiency of the product realization process, while ensuring that the job is done properly. But there are other, more subtle issues to consider.

Proliferation of product variety, coupled with a generally faster pace of new product introduction, has led to shorter product life cycles. With the market life of many electronic products such as personal computers and audio/video equipment now only about eighteen months, a new generation of a product can quickly render an earlier design obsolete so that it can no longer compete. Mechanical products face less extreme pressures but are similarly affected. Under such conditions, it is essential to recover development costs and any product-specific capital equipment costs very quickly. Another implication of this shorter product life cycle is that production equipment must be more flexible; one cannot assume that machinery will be used to make only one product type throughout its useful life.

The increased cost of money has directed new attention to inventory reduction. Instead of relying on storage of a sufficient quantity of components and finished goods to meet all demands, manufacturers must increasingly rely on careful synchronization of supply and demand of these goods. An important consequence of this shift is a tighter coupling of the processes, which in turn implies a need for much better management of a system that is becoming increasingly complicated. Since inventories were originally intended to buffer, and therefore isolate, the effects of disruptions in one step of the production sequence, it is now necessary to contend with "ripple effects" that a disruption can produce throughout the manufacturing sequence.

It goes without saying that the availability of ever more powerful and inexpensive computers has permanently changed the way we do business. This trend will certainly continue. Indeed, computer technology and the system options that it permits will soon dominate the information aspects of manufacturing.

Although most of these trends and issues are well recognized by industry today, it is important to look further ahead to changes that will alter the basic foundations of competitive manufacturing in the next century, and to seek answers to deeper questions. For example, how can the United States compete with countries that have abundant low-cost labor and are also aggressively developing and acquiring advanced technology?

The first requirement is that we accept that changes in manufacturing technology are inevitable. Instead of resisting these changes as we have tended to do in the past, we must find ways to take advantage of them. In terms of the directions for research, this means that we must investigate those technologies that can operate effectively in a changing environment. In more human terms, it means that we must emphasize the kind of education that prepares people for changing roles.

Second, we must understand the necessity of relying comparatively less on experience and more on sound theory. The ability to apply trial-and-error learning to tune the performance of manufacturing systems becomes almost useless in an environment in which changes occur faster than the lessons can be learned. There is now a greater need for formal predictive methodology based on understanding of cause and effect. This methodology can be expressed in a variety of forms: equations, mathematical models, simulations, algorithms, approximations, etc. Of course, a good deal of such methodology already exists, but the practices of industry tend to place greater reliance on experience-based knowledge than on theory-based knowledge. This difference is due in part to the failure of practitioners to familiarize themselves with the analytical tools that are available. In part it is due to a failure of the research community to develop the kinds of tools that are needed and to put them into a usable form.

Although many of the computer aids assist individuals in doing their separate jobs, and sometimes in facilitating communication among them, they generally fail to provide the kind of integrated system that is needed to trace the effects of choices made in one arena of manufacturing upon another. In part it is due to a paucity of sound data on the operating parameters and performance characteristics of complete systems. In part it is because the present situation can be described as an ad hoc collection of programs that were designed separately, each tailored to specific requirements but incapable of functioning as a unit. A wide variety of software tools is needed to handle the information aspects of manufacturing so that alternative choices for actions can be assessed. These tools must extend the scope of concerns beyond traditional boundaries. Moreover, they must be capable of functioning together as a coherent system. This cross-disciplinary integration is not likely to occur, however, without deliberate, focused effort.

Another extremely important guiding principle for research is that we must generate reusable results having broad applicability. The best examples of advanced manufacturing systems that have been commercially developed, including those described in this volume, are tuned to the specific set of conditions in a single plant. Although these systems may provide great benefits to the companies who own them, there are few transferable benefits to the next company wanting to do something similar or even to the same company in another plant. In effect, each new system development project starts over from the beginning. If we are to have the kind of

impact we desire on the whole of discrete manufacturing practice, we must find generic solutions that can be applied in many circumstances.

If future research is to develop the needed tools for design and control of integrated manufacturing, it must emphasize three critical points: (1) more direct industry involvement, so that the work will meet their needs and will be in a usable form; (2) a broader "systems" view, with specific attention to interfaces, so that pieces will not have an isolated utility; and (3) greater rigor, including better performance data, so that we can be certain of what we know and can apply the knowledge to different situations. Applied to specific cases, these principles suggest that research proposals lacking in these characteristics may be less worthy of support than those that do address them.

These generalizations may appear to state no more than what was, or should have been, obvious for years. However, there is a new urgency to the need as we attempt to direct the relatively scarce resources available for research in manufacturing to those key issues on which our future competitiveness will rely. If the views expressed in this volume are a true measure, a substantial number of people are apparently ready to work together to consider collectively where the priorities lie.

Each reader must contemplate the specific conclusions to be drawn. The papers collected in this volume, representing the thoughts of some of the leading experts in various disciplines related to discrete manufacturing, can provide a stimulating beginning. However, there is one additional point that should be borne in mind. The message of this volume is contained in the totality of what is presented, not the separate pieces. Therefore, readers would be well advised to resist the natural tendency to consider in detail only those sections that best match their own interests. As we search for the right path to a future of sustained prosperity in manufacturing, our best hope for collective wisdom lies in open-minded consideration of all of the points of view represented here.

REFERENCES

American Society of Mechanical Engineers (ASME). 1986. Goals and Priorities for Research in Engineering Design. New York: ASME.

National Research Council (NRC). 1984. Computer Integration of Engineering Design and Production. Committee on the CAD/CAM Interface, Manufacturing Studies Board. Washington, D.C.: National Academy Press.

National Research Council (NRC). 1986. Toward a New Era in U.S. Manufacturing: The Need for a National Vision. Manufacturing Studies Board. Washington, D.C.: National Academy Press.

Thomas, Michael E. 1983. A Workshop on Research Directions in Industrial Engineering, Georgia Institute of Technology, Atlanta.

Volz, Richard A., and Arch W. Naylor. 1985. Workshop on Manufacturing Systems Integration, University of Michigan, Ann Arbor.

MANUFACTURING SYSTEMS: MEETING THE COMPETITIVE CHALLENGE

Erich Bloch and Kathy Prager Conrad

The current heightened interest in manufacturing is long overdue. It follows a period during which manufacturing was neglected, often being considered merely a necessary task. Because manufacturing has not been viewed as an intellectual challenge, its intellectual base has not expanded as the products being manufactured have increasingly become more complex.

The manufacturing enterprise is undergoing fundamental change. It both uses and produces sophisticated technology. It is as much a software system as it is a hardware system. The human-factors aspect of manufacturing is also gaining in importance. With this expanding complexity, we need to look at manufacturing as an integrated system and optimize it as a total process, from product design to marketing, instead of suboptimizing on a tool-by-tool or machine-by-machine basis. We must recognize that manufacturing increasingly is driven by advances in science.

Several articles in the recent literature highlight the problem. A recent publication of the Manufacturing Studies Board of the National Research Council noted that "There is substantial evidence that many U.S. manufacturers have neglected the manufacturing function, have overemphasized product development at the expense of process improvements, and have not begun to make the adjustments that will be necessary to be competitive" (National Research Council, 1986, p. 2).

Jaikumar (1986) has analyzed the relative levels of use of flexible manufacturing systems by U.S. manufacturers and their Japanese competitors and has noted that Japanese use substantially exceeds that of comparable U.S. systems. He concludes that this is because U.S. companies are buying sophisticated hardware but using it to replace existing tools rather than integrating it into a new manufacturing system or providing a new approach.

Perhaps the most telling comment about the understanding of manufacturing is the following anecdote recounted in the *Japan Times:* "Official data shows that U.S. telephones are likely to cause trouble three to ten times more than those made in Japan. When [the author] told this to a U.S. Congressman, he said 'Then all we have to do is increase the number of repairmen' " (Karatsu, 1987, B1). Unfortunately, this response is not so different from that of many industry managers we have known over the years. It does not represent an atypical attitude.

In fact, many of the problems for the manufacturing sector started when maintenance and servicing of the product were established as separate profit centers, creating a unit that is expected to be evaluated like other profit centers and to provide a return on investment. Making a profit on product design errors or on the lack of product reliability does not provide a strong incentive to improve customer satisfaction.

Just at the time that manufacturing was being given low priority in this country, the world economy was undergoing dramatic changes. Global competition has now intensified to the extent that we are being challenged on every front. This competition is not just for commercial markets. In basic research and even in education, areas in which the United States has long been the unquestioned leader, our competitors have made vast inroads.

Competitor nations are now recognizing the validity of a strategy that has long been followed by the United States, namely, that top research universities can be the genesis for R&D-intensive companies. The commitment of these countries to strengthen their basic research in manufacturing could be especially significant, since our basic research efforts in this area have been minimal until now.

We see this approach being taken in both Japan and the Federal Republic of Germany. Japan is planning a new mechanism to breed strong basic researchers—the so-called technopolis project (The Economist, 1986). On various sites, each called a technopolis, a university, local government, and industry are being brought together. The technopolis is intended to promote basic research, with the goal of founding new high-technology companies. About 18 of these new units have been approved. Similarly, the German government has just established a major new center for production technology (The Economist, 1986). This center will be jointly used by the Technical University of Berlin and the Fraunhofer Gesellschaft. The emphasis will be on computer-integrated manufacturing, with state-of-the-art equipment and strong industry involvement. The investment in this single center will be the equivalent of about $67 million—more than the total committed by the National Science Foundation (NSF) to the Engineering Research Centers that are concentrating on manufacturing technology.

In the face of these efforts and our general lack of interest in manufacturing, the United States has adopted a primarily defensive strategy. We have blamed unfair trade practices, the overvalued dollar, and lower wage scales of other nations. Instead of criticizing our competitors and trying to create a level playing field—not unimportant but not sufficient—we must examine our fundamental approach to the problem and adopt a tough offensive strategy.

Historically, our comparative advantage has been knowledge and our ability to develop new bases of knowledge rapidly through our universities. In manufacturing, however, we are not developing sufficient new insights to support the next generation of production processes. Neither are we deploying the knowledge and technology that we have in a timely and appropriate way.

Many things need to be done to change this situation. One obvious need is for more interaction among universities and industries. This is essential because technology transfer is really knowledge transfer. Bringing university researchers who have ideas together with industry colleagues who can use those ideas to solve problems will ultimately lead to new or improved products and processes. But this is not happening to a sufficient extent in the manufacturing arena, partly because of several more fundamental problems facing both universities and industries.

Consider the universities first. As noted earlier, manufacturing has held little attraction in our universities over the past decades. Manufacturing programs have only recently begun to develop, with less than six accredited manufacturing engineering programs at the bachelor's and master's levels in U.S. engineering schools. Although a large number of manufacturing-oriented programs have sprung up in the past few years, few have gained the accreditation that gives them the stature necessary for long-term continuity. Unfortunately, the mind-set that fails to recognize manufacturing as a legitimate academic field persists, and despite the emergence of a few strong programs, there is no universal recognition of the need.

In fact, many university researchers are hesitant to tackle the critical issues. They are skeptical of the intellectual challenge and concerned that the relevance to industrial needs might be misunderstood by their peers. This situation is particularly disconcerting because basic research is needed in many areas. Examples include the integration of information into manufacturing systems, particularly in achieving designs that optimize the manufacturability of the product, process planning, and the innovative use of new materials. These are good topics for university researchers. They should be considered valid areas of intellectual inquiry by the academic community-at-large, not just the handful of schools now making efforts in these fields. To make progress, the university system must recognize the value of investigating relevant problems and en-

courage young faculty members to take on these important research questions. This is not likely to happen overnight or on its own; catalysts for change are needed.

The National Science Foundation is increasingly playing such a catalytic role. In addition to encouraging individual investigators in this direction, it also supports multi-investigator efforts through the Engineering Research Centers, the Materials Centers and Groups, and now the Basic Science and Technology Centers. These activities provide opportunities for students and faculty to move outside the boundaries of the traditional disciplines and work with people having different perspectives. In multidisciplinary areas such as manufacturing, which cut across the disciplinary structure of universities, this is especially critical. University and industry researchers are brought together in an interactive setting, developing partnerships that are based on more than money to support the research. Further interaction depends on the quality and continuity of these relationships.

The attractiveness of these programs to researchers interested in manufacturing is evident from the high proportion of the Engineering Research Center proposals that have addressed manufacturing research areas and from the high number of these proposals that have received funding. Although such developments are encouraging, clearly much more needs to be done.

States are playing an increasingly active role in bridging the gaps between universities and industries. Currently, 27 states have started programs to develop university-industry research partnerships. In many cases, the NSF Engineering Research Center program has been used as a model.

The problems that need to be solved are not just in the universities. The industrial sector has both an immediate and a long-term challenge. Even in high-technology sectors where the United States once held an unqualified world lead, its position is now threatened. In some cases, the gravity of the situation is drawing industries together, and new offensive strategies are being developed.

An example of this can be seen in the semiconductor industry. There is widespread agreement that steps must be taken now to restore the U.S. edge in semiconductor technology and manufacturing. A recent survey of the Semiconductor Research Corporation's member companies suggests that the U.S. lead in several critical semiconductor areas is expected to drop substantially by 1991 (Burger, 1986). In response to concern for the declining competitiveness of the semiconductor industry, the Department of Defense convened a special panel to examine the situation. Its recently released report calls for the establishment of a manufacturing consortium that would concentrate on advanced equipment and process research, providing a resource that member companies could use in the marketplace to further their competitive advantage (U.S. Department of Defense, 1987).

It is clear, however, that creating a new research consortium will not solve all of the semiconductor industry's problems. Although new knowledge and technology will be generated and the interactions that are necessary for progress will be improved, these institutional and research tools alone will not be sufficient. More basic challenges face U.S. companies: to learn to think and work with entire systems and to adopt an overriding concern for the quality of the product.

In addressing the issue of quality, several myths about the relationship between quality and productivity need to be dispelled:

• The first is that quality is an extra, a luxury desirable to the extent it is feasible. In fact, quality is an essential requirement and does not come in increments—low quality is not an acceptable or meaningful standard.
• The second is that quality is not quantifiable—that it is a subjective judgment by the consumer. It is, in fact, precisely measurable.
• The third is that quality is expensive. This is a short-term management attitude that ignores the long-term savings achieved through quality management.
• Finally, there are many misconceptions about the causes of poor quality. By ignoring measures such as error rates, scrap, and other indicators of the integrity of the design and manufacturing process, this myth attributes poor quality to lazy workers, labor-management disputes, and other environmental factors.

The reality is that there is a strong correlation between quality and productivity. By using a quality management system, companies can make better products more efficiently, thereby increasing their productivity and their ability to compete. Commitment to such a system requires a long-term outlook.

U.S. managers have been preoccupied, however, with short-term results and quarterly reports. As a result, they have lost sight of long-term goals and the strategies needed to achieve them. For example, when labor costs in this nation rose, many companies turned to overseas production without proper concern for quality, responsiveness, or the impact on the total enterprise, its product line, and the servicing of its products. In many cases, fixing the shop at home through new processes and approaches would have been a more viable solution for the long term.

Any improvement in quality must be achieved by people, and this will require changes in the attitudes and practices of today's work force and management and in the way we educate the scientists, engineers, managers, and production workers of tomorrow. New linkages must be established, not only between universities and industries but also among the various components of individual institutions. In our companies the research people need to talk to designers, assembly workers need to talk to system engineers, and so forth. In the universities, schools of business should be addressing important manufacturing issues from a broader perspective than finance or management. Curriculum changes are needed in the business schools and the colleges of engineering and science. Students and faculty of the various disciplines must become familiar with each other's language and tools.

These issues are indicative of one of the main problems in U.S. manufacturing—a lack of contact and communication among the different sectors and organizations that have a stake. This is as true in the government as it is in any other sector. One effort to foster such interaction is the Manufacturing Forum, to be established by the National Academy of Engineering in cooperation with the Office of Science and Technology Policy and the National Science Foundation. The Forum will bring together cabinet- and subcabinet-level government executives, chief executive officers of several manufacturing companies, university presidents, and representatives of labor. The objective of the Forum is to provide an environment that will promote open discussion of manufacturing issues. It is hoped that this focus on an important national problem will be productive and will facilitate cooperative initiatives among the participants that otherwise might not occur.

Such activities show that we are making progress. Additional effort is needed, however, to speed the positive action that is already occurring and to encourage further initiatives. We need to establish a sense of national priority in manufacturing and manufacturing research. Manufacturing should be at the top of the agenda in government, corporate board rooms, factory floors, and universities across the country. Working together, we can trade on our strengths—the inventiveness and dedication of our people—to enhance our competitive position.

REFERENCES

Burger, R. 1986. Semiconductor Research Corporation Newsletter 12:4–8.

The Economist. 1986. Japanese research: More sci than tech. Nov. 22, p.94.

Jaikumar, R. 1986. Postindustrial Manufacturing. Harvard Business Review Nov./Dec.:69–76.

Karatsu, H. 1987. Healthy economies are founded on sound manufacturing bases. Japan Times, Jan. 22, p.B1.

National Research Council (NRC). 1986. Toward a New Era in U.S. Manufacturing: The Need for a National Vision. Manufacturing Studies Board. Washington, D.C.: National Academy Press.

U.S. Department of Defense. 1987. Report of Defense Science Board Task Force on Defense Semiconductor Dependency. Washington, D.C.

DESIGN AND ANALYSIS OF INTEGRATED ELECTRONICS MANUFACTURING SYSTEMS

Laurence C. Seifert

ABSTRACT   AT&T's reaffirmation of its commitment to manufacturing excellence has resulted in a number of productivity improvement programs. Processes are now designed to survive model changes, avoid suboptimal use of robotics and computing technology, and deal with culture factors. It was first necessary to focus resources and management attention on the commitment to manufacturing. Second, a methodology was evolved that mandates, in order of priority: (1) up-front systems analysis and engineering, (2) application of "suprahuman" processing facilities along with automation of material handling facilities, (3) automation of information flows, and (4) reduction in labor through development of physical automation. Projects emphasize implementation management and dealing with the human tendency to resist change. A set of needs and opportunities for the U.S. technical community to achieve productivity improvements is described. It is AT&T's experience that the current capabilities of technology, the tools that support it, and the capabilities of the engineers who must implement it are more than sufficient to meet overall goals. Simply stated, present capability is not being widely applied across U.S. manufacturing operations.

INTRODUCTION

The past five years have been a difficult, yet exciting, period for the manufacturing segment of American industry. At AT&T, in addition to the more publicized divestiture of the local Bell Telephone Operating Companies, we have experienced a revival in our commitment to manufacturing excellence. This renaissance is fueled by the reaffirmation of the strategic link between manufacturing and distribution productivity and the achievement of our corporate goals. We firmly believe that manufacturing is a principal vehicle in bringing the benefits of new technology to market. This paper, however, is intended neither as a chronicle of the events that are reshaping manufacturing nor as a dissertation on the causes. It deals instead with what we have learned from our activities in manufacturing productivity improvement. It also de-scribes the course we have set for ourselves, and it identifies problems and opportunities for furthering productivity gains.

AT&T has conducted productivity improvement programs at various manufacturing facilities. This experience is the base from which we have established new directions for further advances in manufacturing productivity. We have a vision for our manufacturing business that we continue to refine. Specific productivity improvement programs and related R&D activities are being driven to achieve this vision.

Quality and simplicity are the cornerstones of our programs. The key factors for realizing the programs' full benefits are

• the commitment of all involved personnel and levels of management, and
• designing and implementing total manufacturing and distribution systems and their linkages with other systems.

Manufacturing operations, which once were the Western Electric Company, Inc., are now deployed as integral parts of the AT&T business groups. A close community of manufacturing management is maintained to facilitate intergroup operations and to leverage R&D and other corporate activities. This is accomplished through a corporate oversight function.

AT&T provides information movement and management (IM&M) services and the systems and products that support these services. This paper concentrates on the systems and products that are sold to customers and are the basis for the IM&M services.

AT&T is a vertically integrated company in the sense that it employs a broad range of internal functional expertise in the product realization processes. Figure 1 shows the grouping of AT&T's manufacturing and distribution operations and their underlying technologies. Table 1 describes the range of in-house R&D and production activities. Basic research is not included in the diagram for simplicity, although we maintain our commitment to this essential activity.

This structure is the focus of our programs in manufacturing productivity improvement. The following functions are contained under the heading of manufacturing:

• Process design and engineering
• Translation of product design information into manufacturing information
• Production planning, scheduling, and control
• Incoming material control
• Material ordering and stocking
• Product fabrication, assembly, and testing
• Product repair
• Quality control
• Product and process productivity improvement
• Manufacturing information management

We consider manufacturing and distribution as a single business activity. This approach integrates the following functions with those listed above:

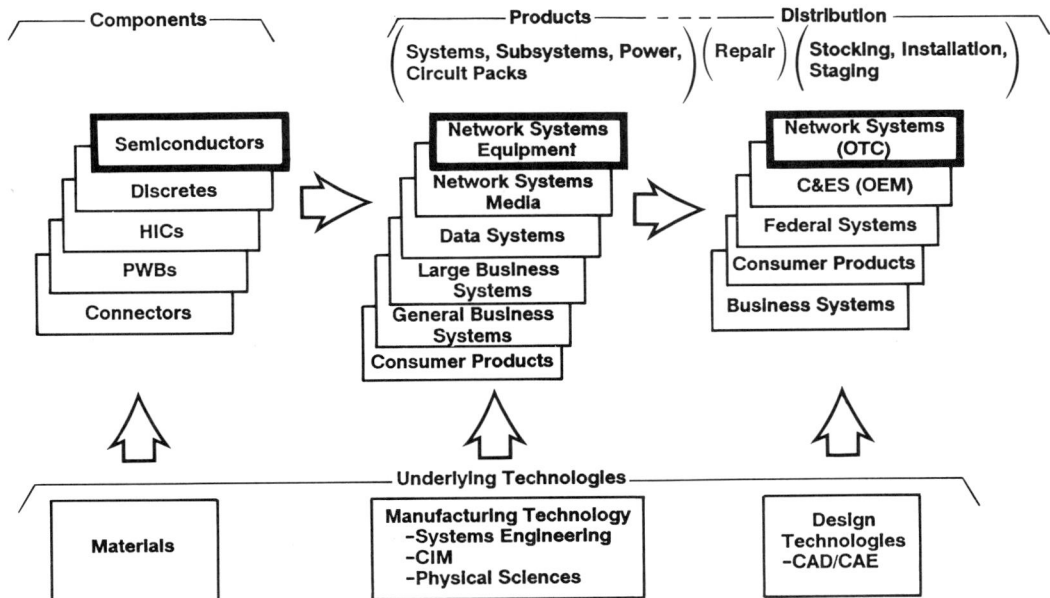

FIGURE 1   AT&T products business groups.

TABLE 1   R&D and Manufacturing Activities

Category	AT&T R&D	Production AT&T	Contract	Purchase
Common Technologies				
Materials	X		X	X
Product CAD	X			X
Manufacturing technologies				
Systems engineering	X			
Systems integration	X	X		
CIM software	X	X		X
Production facilities	X	X	X	X
Components				
Integrated circuits	X	X		X
Discrete devices	X	X		X
Interconnection technologies	X	X		X
Systems and Products				
Media (light guide, cable, etc.)	X	X		X
Electronic/photonic systems	X	X		X

- Management of orders
- Staging of customer system and materials
- Stocking of finished goods
- Installation and customer service

This model also employs product-focused operations, as compared with functional operations, which are minimized. Terms such as "focused factories" are used to describe these operations. They will eventually result in a merger of manufacturing and distribution operations into product family operations that are singularly managed.

The balance of this paper deals with our experiences with productivity improvement programs and the directions we have set for design of manufacturing processes and for related R&D programs.

A LEARNING EXPERIENCE

Our successes have exceeded our failures, but understanding both is essential. Failure mode analysis is just as valuable in acquiring knowledge about integrated manufacturing systems as it is in achieving yield improvements or in refining chemical reactions.

The following generalizations, culled from our learning experiences, are presented as major examples of what not to do again—or as one answer to the question many managers ask, "Why does computer-integrated manufacturing fail?"

Do Not Accept Process Performance As It Is

Yield improvement, normally considered an element of quality improvement, almost always offers the most significant opportunity for improving process productivity. Design for manufacturability (DFM) and its companion, failure mode analysis (FMA), have not always been the primary targets of productivity improvement projects. Yield-quality-FMA programs—that is, a total quality control (TQC) program, must be implemented before new facilities or new automation capabilities can provide the expected performance benefits. When TQC has been addressed, performance usually has exceeded expectations, and additional production capability has been realized. If capacity planning is not done on the basis of productivity goals, it is likely to deploy more capacity than is needed. This is fine if marketing is able to sell the added products. If not, this extra capacity, once deployed, carries an ongoing earnings penalty—a cost that detracts from financial performance of the processes.

Do Not Do the Wrong Thing a Bit Faster

Physical automation, both specific "hard" automation and more flexible robotics, has been implemented on occasion with only marginal improvements. An example of this is the use of robotic workstations to insert "odd-form" electronic components into printed wiring boards, when the electronic circuit functions could well be performed by "standard" high-speed machine-insertable components.

Figure 2 shows the relative cost of assembly for three types of components using four different methods for inserting the components. First is hand assembly, whereby a single operator assembles all the components for a printed wiring board. The second method, called progressive hand assembly, consists of a line of operators, each assembling repetitively a limited set of components. Industrial engineering studies have demonstrated that an individual cannot ef-

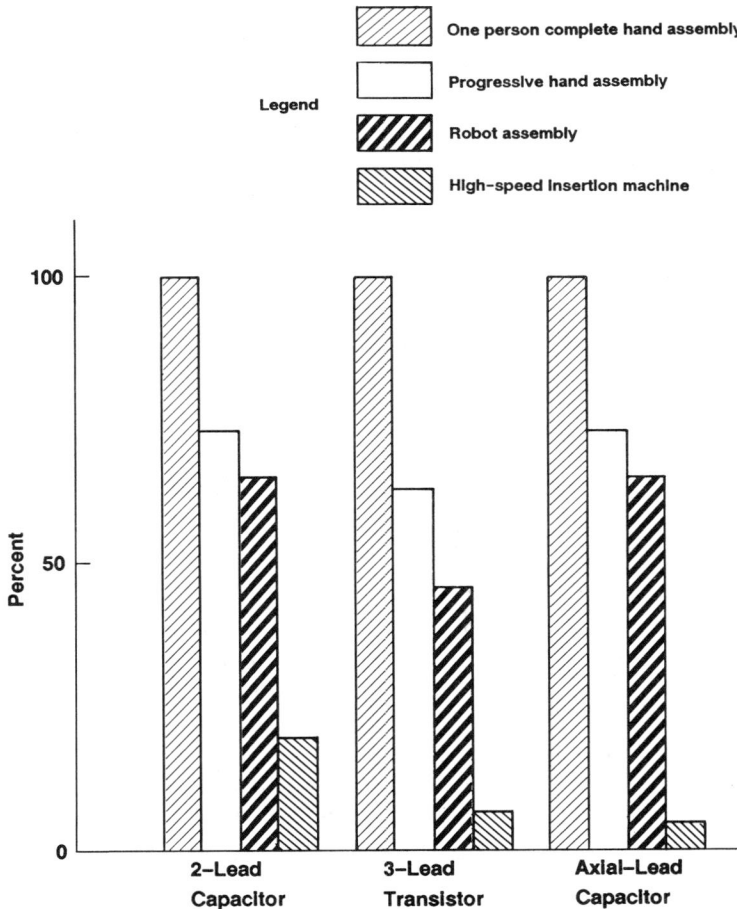

FIGURE 2 Relative complete cost of component assembly.

fectively, with high-quality performance, assemble a large number of components on a repetitive basis. The third method is the use of a robot to perform the task. The cost of assembly by robots not only is lower but also yields higher quality by lessening the "human error" factor. The fourth method uses high-speed machines configured specifically for the task.

Normally, odd-form components can be installed only by the first three methods, with the last method being limited to components conforming to certain physical parameters. However, we have found few cases in which a circuit cannot be *designed* with components that are compatible with the last method of assembly. Examples of exceptions to this generalization are analog and high-voltage components. The use of high-speed dedicated machines results in much lower assembly costs than can be achieved by the first two methods, a difference that is greater than the differential in labor rates between the United States and most Asian countries. Preferred component use offers not only improvements in assembly costs but also additional benefits realized from material management, parts stocking, volume purchasing, test program reuse, and fewer quality and reliability engineering activities.

A strong DFM effort between product designers and manufacturing engineers, especially for component engineering in the manufacture of electronic circuit packs, has greater leverage in reducing overall costs than does automation aimed at simply reducing assembly labor cost.

Surviving a Model Change

Implementing major automation projects only to have the manufacturing process not survive a model change is the most costly expenditure of all. This is why flexible manufacturing is everyone's objective. It is essential, however, that one clearly understands the circumstances under which

hard or flexible capital facilities are desirable and when more manual alternatives are appropriate.

Avoid Suboptimal Use of Computing Technology

Applying computing technology to various operations mandates a careful coordination of all aspects of information flow. The suboptimal use of computing technology often leads to confusion, requiring engineering and material management personnel to adjudicate differences.

An example illustrates this situation. In one instance, product information came to the shop floor by three separate paths, with different functional groups independently "adding value" to the information. Further, there were 14 manual information translation points between the various information systems that served the three paths. It was obvious why the information on the shop floor was inaccurate.

The Culture Phenomenon

Adequately dealing with people and their natural reluctance to accept change is critical. How many of us have heard comments like the following:

"There's nothing wrong with our manufacturing processes; the problem is the product designers."

"We can't afford to spend money on productivity programs."

"We already have too many engineers."

"We simply need to get labor rates down."

"If only the product forecasts were more accurate."

Many believe that the most difficult problem to overcome is the propensity for operating personnel to act on their own authority, for engineers not to listen to problems, and for management to fear and resist change. As W. Edwards Deming remarked in a recent seminar, "If we know what to do, why don't we do it?" We can design and implement the best manufacturing processes and facilities, but if its users are not "buying in," we have wasted time and resources.

Overall, the most important element in achieving an improvement is fully comprehending the problem. After all, if there were no problems, there would be no opportunities for improvements. But a superficial understanding of a problem will lead to solutions that generate still more problems.

DIRECTIONS

A Corporate Focus

Our current approach to achieving ongoing manufacturing excellence has a number of elements. Foremost among these elements is a set of activities that focuses resources and the attention of personnel on manufacturing, thereby demonstrating a commitment to doing what is necessary to bring continuous improvement to the manufacturing operations. Executive attention and support, the allocation of necessary resources, including manufacturing engineering and central R&D, and the establishment of performance goals are key factors for achieving success. AT&T has put in place the following activities to accomplish this.

A Manufacturing Technology Board

Each business group has a manufacturing technology board that is made up of the chief engineers of each of our manufacturing facilities, representative product development directors, and a manufacturing R&D director. The objectives of these boards are to develop plans for improving the productivity of the manufacturing operations, for overseeing the programs, and for ensuring commonality among operations. Each of the directors typically oversees each specific program—for example, information automation or surface mount technology implementation.

A Corporate Manufacturing Officer

A corporate manufacturing officer, reporting directly to the president of the corporation, has been designated to facilitate manufacturing planning. This individual is charged with establishing and implementing a corporate manufacturing plan, which must be approved annually by the highest levels of the corporation. Organizations supporting this office consist of functional planning groups for various aspects of manufacturing operations, the manufacturing R&D operations, and the environmental control staff.

A Manufacturing and Distribution Council

A manufacturing and distribution council, consisting of all corporate officers who have direct manufacturing or distribution responsibilities and chaired by the corporate manufacturing executive, has been established. The goal of the council is to facilitate and oversee the development and implementation of important plans for all manufacturing and distribution functions. The council's objective is to ensure that AT&T's critical manufacturing and distribution resources are optimized for the benefit of AT&T as a whole.

Internal Manufacturing R&D Capabilities

The internal manufacturing R&D capability was started 30 years ago. It has recently been supplemented with the Manufacturing Development Center, a capability for production facilities systems integration and replication, and with a significant commitment of resources by AT&T Bell Laboratories for computer-integrated manufacturing (CIM) systems development.

Manufacturing Systems Eng (MSE)

MSE – Process Eng

Characterization, design *, and engineering of manufacturing processes.

Design for Manufacturability (DFM)

Information Mechanization

Mechanization = Integration ** + Automation

Physical Automation

For:

Suprahuman Processing

Technology/machinery that do what humans cannot do well.

Flow Control

Material Handling, Queue Control

Labor Productivity

Priorities

MSE Process Eng
Design for Manufacture

First

Suprahuman Processing	Then
Flow Control	Then
Information Mechanization	Then
Labor Productivity Automation	Last

* **Emphasis: Customer expectations/simplicity/TQC, with JIT, in-lining, flexibility concepts**

** **Linkages: product CAD, distribution, financial, and other information systems**

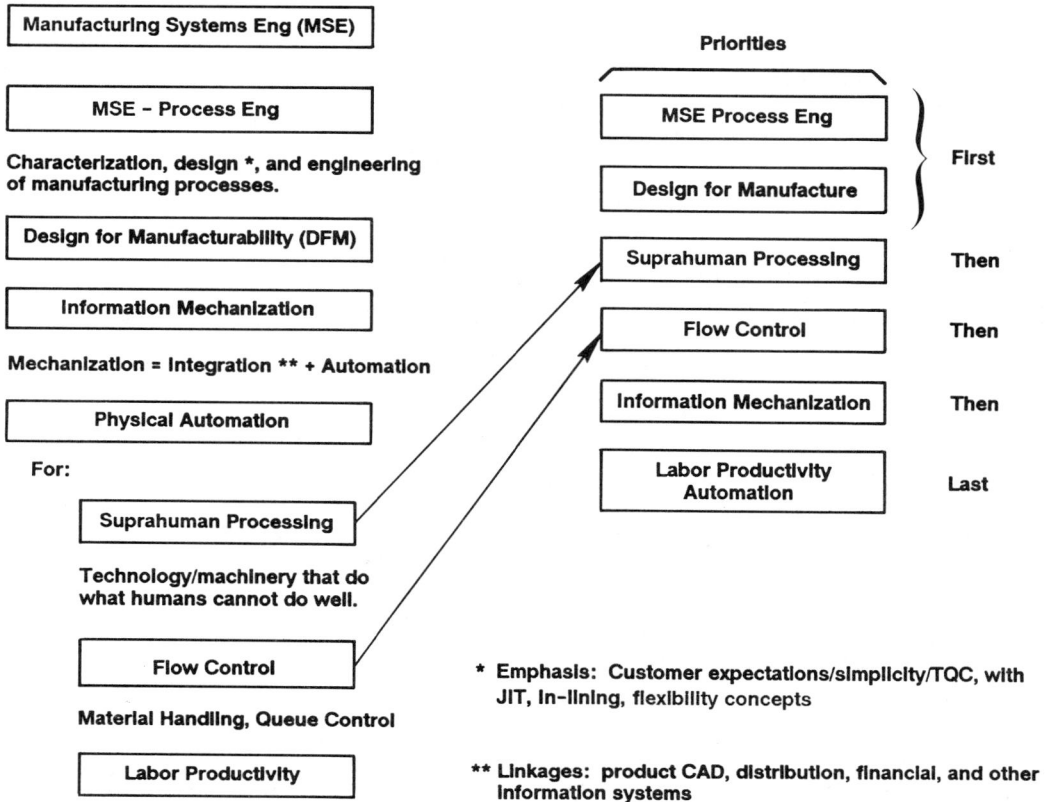

FIGURE 3   Manufacturing productivity realization model.

High-Visibility Projects

Specific high-visibility projects have been established in each business group. These projects, targeted on the group's business priorities, are supported by central R&D personnel. The purpose of these initiatives is to accelerate productivity improvements in certain operations. The projects also effect a technology transfer, to the operations engineering staffs, of a number of technologies, with recently increased emphasis on manufacturing systems engineering disciplines.

Projects that demonstrate significant improvement serve as an example—an existence proof—of what is possible. They stimulate ongoing work and other projects. Care has to be taken, however, to ensure

that this is not just a Hawthorne effect.[1] It is essential that the disciplines necessary for ongoing improvement, as well as the desire for ongoing improvement, be built into the fabric of the operations.

Methodology

The second element of the program has been the establishment of a *methodology* for productivity improvement programs. Previous experiences have led us to follow a sequence of activities, as shown in Figure 3.

[1]The methodological Hawthorne effect is usually defined as the problem arising in field experiments when subjects' knowledge that they are in an experiment alters their behavior from what it would have been without that knowledge. Documentation of this effect came in the classic studies at the Hawthorne Works of Western Electric Co. (Mayo, 1945).

The first principle of the methodology is that redesign and reengineering of the processes must take place *before* an attempt is made to apply either information automation or physical automation. We call this discipline manufacturing systems engineering (MSE). Most benefits come from simplifying systems, improving yields, and reducing the cost of quality.

The techniques of MSE and the tools that support them are based in traditional industrial engineering practice. The current tools, however, are more powerful because of today's computing technology. Central R&D resources have been valuable in developing these new tools, establishing technology transfer procedures, and supporting specific projects.

We have embarked on a major program aimed at fostering the use of systems analysis disciplines and process design and engineering tools. Figure 4 shows current results by category and number of tools that are supported centrally by R&D. The percentage of facilities using these tools might be somewhat misleading, as it was calculated by the number of locations using the tools at least once. Use has been accelerating, and the results are encouraging. An annual technical symposium is held for AT&T manufacturing systems engineers, papers are presented, and much "networking" goes on. Attendance has grown from 35 in 1985, to 85 in 1986, and more than 200 in 1987. Tools developed or supported by the R&D staff are listed in the appendix to this paper (Eckel, 1986).

The second principle of the methodology is that manufacturing is information-intensive. Many believe that mechanization of information transfer through computer technology will result in significant improvements in productivity. Some have succeeded and others have failed to achieve

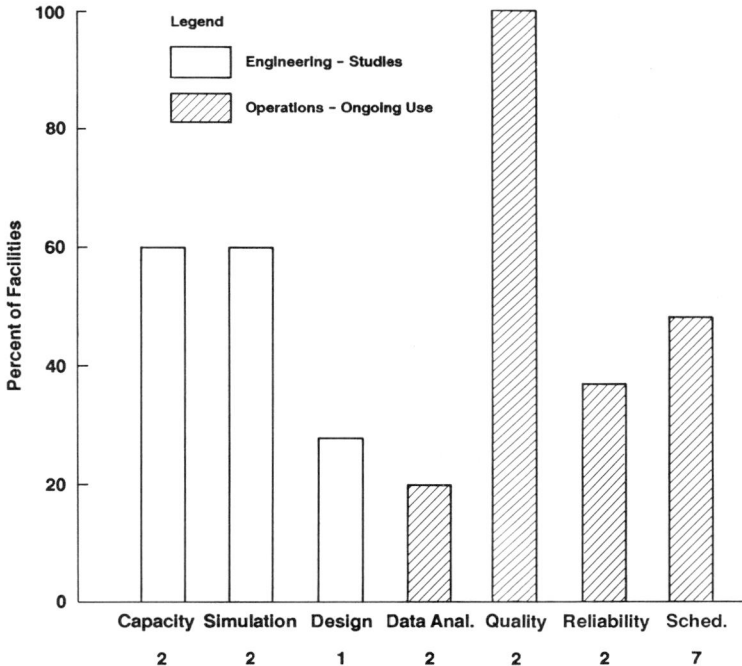

FIGURE 4 Implementation of process engineering software tools at AT&T manufacturing facilities. The number centrally supported is shown beneath each column.

this promise of CIM. CIM can become the vehicle for embedding procedural discipline in operations by reducing the need for manual intervention. Since mechanization equals automation plus integration, the major benefits of information mechanization come from the integration component of that relation.

Providing a customized information system for each operation can be expensive, both for initial implementation and ongoing enhancements. Therefore, AT&T has developed a set of corporately supported systems using modularized open systems software architectures. Systems meeting the requirements of various production processes can be individually configured. They also provide users some flexibility to better adapt the systems to their needs.

Difficulty arises in interfacing the array of various suppliers' production facilities with the system. Most new manufacturing equipment, which once had only program logic controllers (PLCs), now requires increasing amounts of computing power. There are few information or data-base

standards or interface protocols in most manufacturing operations. Many efforts are under way now in this country to standardize the interfaces to obviate this problem, the most notable of which is the manufacturing automation protocol (MAP), championed by General Motors (Jones, 1988).

Figure 5 shows the levels in the computer-integrated manufacturing architecture (CIMA) that AT&T has adopted for internal use. Also shown are the system design requirements for response times by level. These requirements result from traffic studies of information flows in many advanced operations.

Figure 6 is a systems diagram of the family of factory floor systems. The systems family that has been chosen to be supported with a corporate information system, called productivity improvement systems for manufacturing (PRISM), has been initially structured for product assembly and testing operations. The integrated circuit fabrication factories use another, internally developed, family of systems. Figure 6 uses the following acronyms and abbreviations:

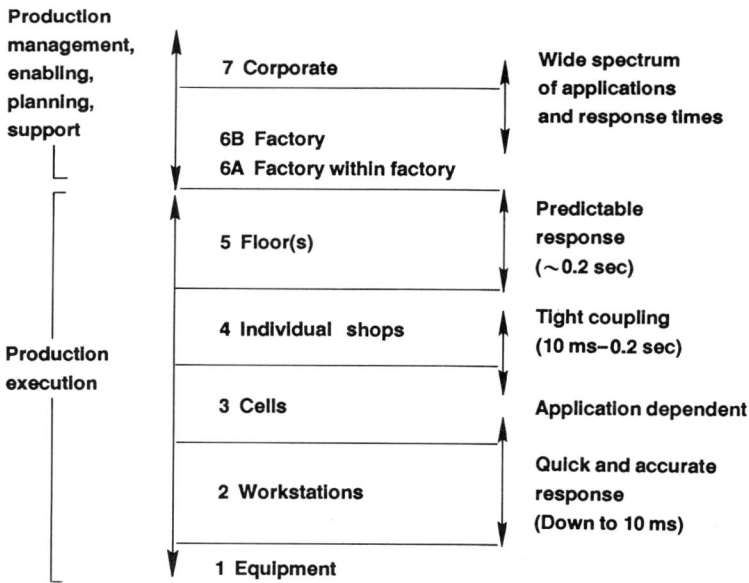

FIGURE 5   CIM systems response time requirements.

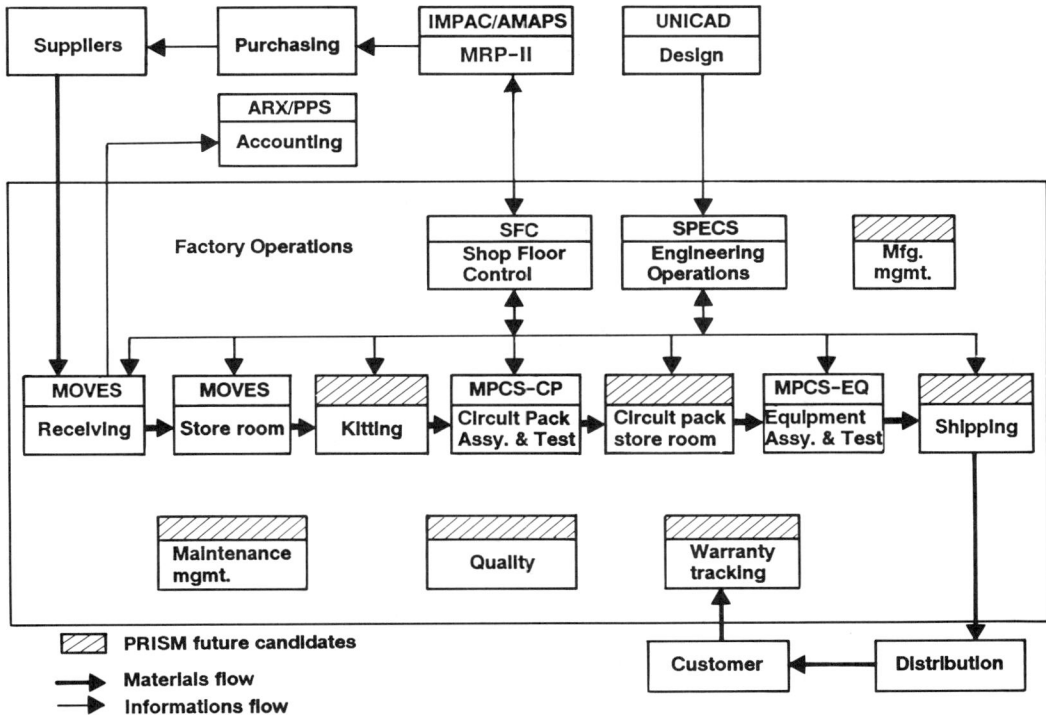

FIGURE 6   PRISM areas of focus.

SFC	Shop Floor Control	AMAPS	Advanced Manufacturing Accounting Production System
SPECS	Synchronized Production Engineering Control System		
MOVES	Materials Operations Velocity System	MRP-II	Manufacturing Resource Planning
MPCS-CP	Manufacturing Process Control System—Circuit Packs	UNICAD	Unified Computer-Aided Design
MPCS-EQ	Manufacturing Process Control System—Equipment		
MPCS-LOT	Manufacturing Process Control System—Lot Processing		
ARX	Accounting Receiving Executive System		
PPS	Planning Procurement System		
IMPAC	Integrated Manufacturing Planning and Control		

For those functions shown without an information system title, local systems are currently in use.

Rather than take the reader through a long discourse on the functions performed by each system, suffice it to say that we have accomplished and implemented several paperless information streams. For example, we are able to generate assembly aids for shop personnel electronically through direct linkage with the product designers' CAD system.

The current status of deployment of these

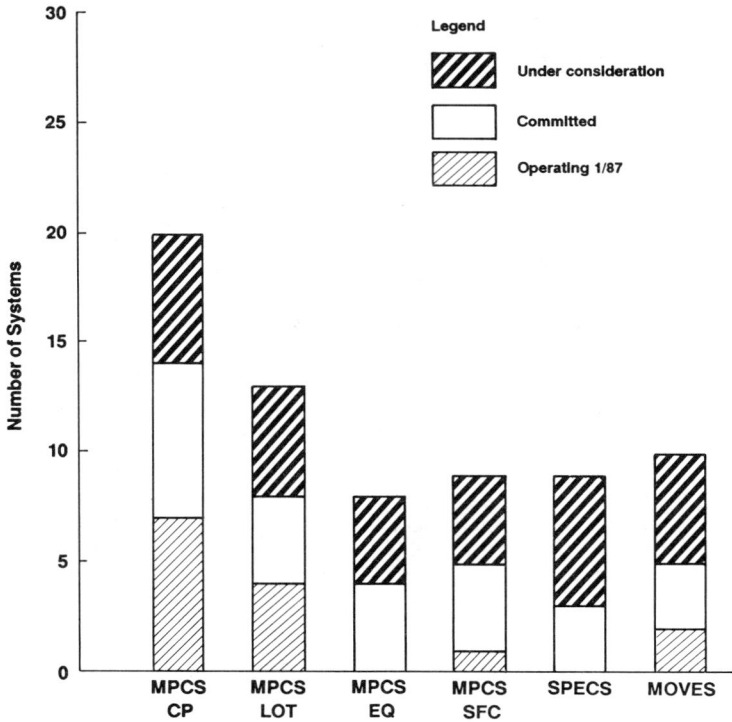

FIGURE 7   PRISM deployment.

systems is shown in Figure 7. Of course, AT&T 3B computers are used, and all software runs under the UNIX operating system. What has been described is a successful CIM program. It is an emulation of the approach AT&T has used for similar systems that we have been providing for telephone companies for more than 15 years. Figure 8 shows the structure of a typical system. Highlights of the approach are these:

• Users are involved in writing system requirements.
• Users are involved in managing development programs and priorities.
• A structured programming development environment is maintained.
• Full commercial documentation, training, and maintenance support, including 24-hour on-call backup, are provided.
• Software is structured to (a) allow unique application configurations; (b) ac-

commodate site-specific programs and interfaces; and (c) provide users with some design space.

We are now in the process of developing real-time versions of several of the engineering and quality analysis tools discussed. When these tools are embedded in PRISM systems, a whole new set of capabilities will be available to local engineering and operating personnel to achieve unprecedented operating efficiencies.

Advances in microcomputing are now making it possible to achieve true real-time process control with no human intervention. Much of our R&D effort is aimed at exploiting this opportunity. We have already built and installed workstations that include all elements of the process control model shown in Figure 9 and are connecting these workstations to PRISM systems. In one example of this procedure, a modification developed for photolithography

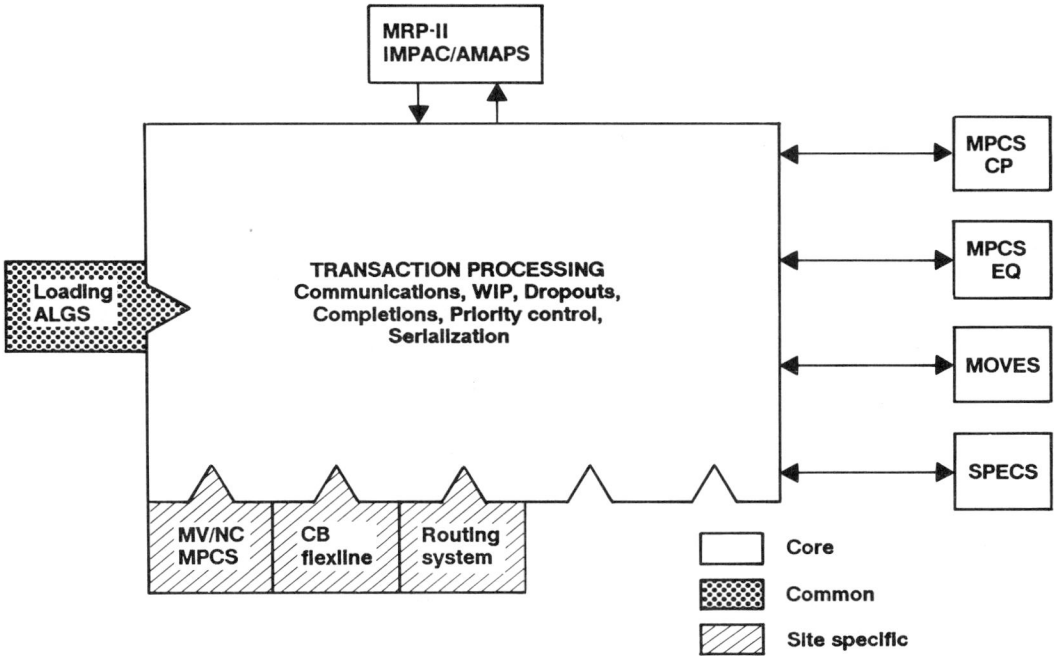

FIGURE 8   PRISM open architecture SFC example.

steppers provides an analysis of processed silicon wafers, knowledge-based analysis, and direct feedback control to a motor-driven platen to adjust for microscopic deviations.

The third principle of the methodology is that physical automation is introduced to replace labor or to facilitate product flow only after completion of the aforementioned product and information system de-

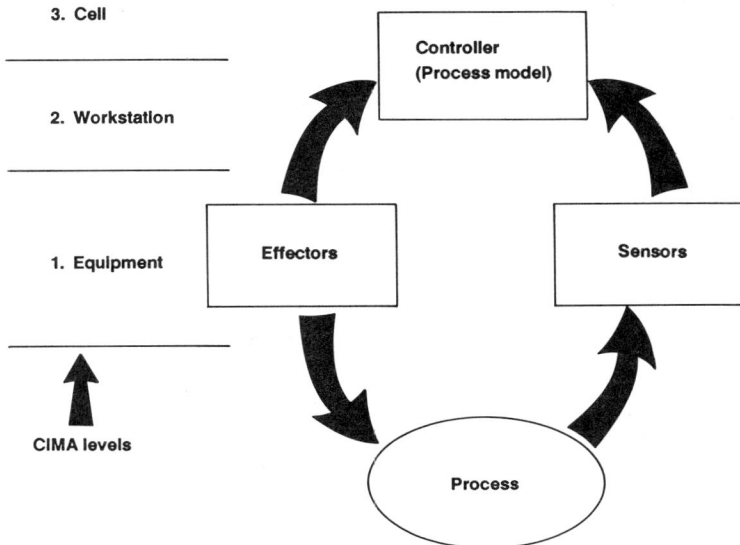

FIGURE 9   Process control model.

signs. Physical automation should be applied first where there is no choice but to use machinery—that is "suprahuman processing." For example, as more surface-mount technology is deployed for the assembly of components to circuit boards, it has been found that high-pin-out IC packages with a pitch of 25 mils or less between leads require assembly by vision-assisted machines for accurate placement. Suprahuman processing, obviously, is the dominant mode in semiconductor wafer fabrication. But much IC packaging continues to be done in a more manual mode. Process discipline and control can be facilitated, as with information flows, by automation of product and material handling facilities. The second choice for automation is for better control over the operation of the process.

After all of the other priorities are satisfied, the final choice for automation is the use of machines for the replacement of manual labor. We all have visions of totally automated processes, the so-called lights-out factory. Great care must be exercised, however, not to generate a system that requires more ongoing engineering labor for product changes and enhancements than has been reduced in the assembly and test operations through automation.

PROGRAM IMPLEMENTATION

The best designs and most accurately focused programs are of no value if they are not implemented or if they are not properly used after implementation. Careful "design for implementation," implementation planning and management, and user "buy-in" are required for successful programs to improve manufacturing productivity. The most successful approach has been the one in which the final user "project manages" the program. This approach leads to more complete consideration of how new operating disciplines affect the users. People are more likely to follow design intentions and

deal with change if they are a party to generating the change.

There has been a great deal of debate over whether it is better to rearrange a manufacturing process in an existing operation, while production continues, or to build an all-new process in a new location, the "greenfield approach." The advantage of the greenfield approach is that production is not disrupted. Verification procedures can be accomplished with limited external variables. The disadvantages are that the new process is not tested in real production situations. Furthermore, there are additional costs for product samples and duplicate facilities. The greenfield approach is best used for production lines for new products. The advantages of rearranging existing processes include reuse of existing facilities and, probably most importantly, the availability of an analytical characterization of process performance. Rearrangement while in production requires a much closer working relationship with the operating personnel. In this situation, the handoff of the new process designs and technology is evolutionary, and the technology transfer to the operating personnel is continuous.

SUCCESSES

Projects that followed AT&T's manufacturing productivity realization model (Figure 3) have yielded remarkable results. Two such examples are the power unit shop at the AT&T Denver Works and the entire operation of the AT&T Oklahoma City Works. Both factories are product assembly and test operations and can be characterized by the model shown in Figure 10. These are "focused factories" in that they perform all manufacturing functions described earlier, with the exception of component fabrication.

These are large factories and, for certain product families, are separately managed as "factories within a factory." This ap-

Logical process:

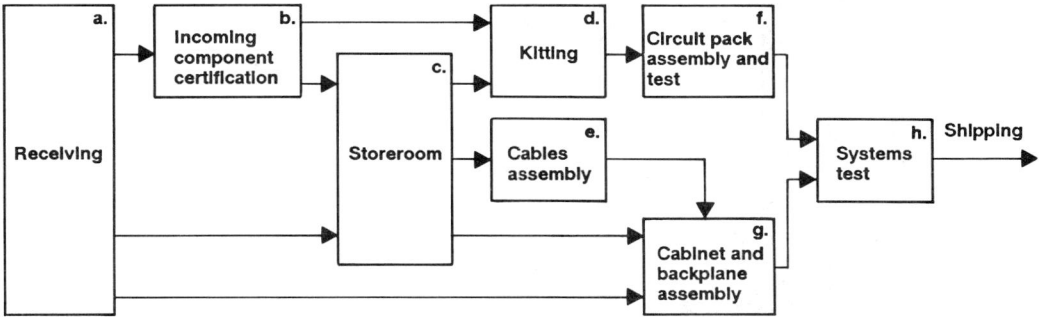

FIGURE 10   Assembly and test operations model.

proach has been likened to shopping malls, where certain common services (e.g., security, heat, and light) are shared on a billable basis, but each store unit is responsible for its own business performance.

The Denver power unit shop employed a few dozen people and represented only a portion of Denver's operations (see block "g" of Figure 10). Since it had always presented problems, it was identified as the pilot for the Denver Works improvement program. The operations were analyzed and characterized, and it was decided to reengineer the process to an in-line, just-in-time, manual *kanban* operation. The engineers concerned set up desks on the shop floor and involved shop personnel in design and implementation phases of the project. They stayed with the shop until everyone was comfortable that the performance of the new process met objectives. No physical or information automation was involved. The results are shown in Figure 11.

Oklahoma City Works employs approximately 6,500 people. The manufacturing operation for the digital switching system

Power Unit Assembly Shop
Application of Just–In–Time (JIT) and Total Quality Control (TQC) Techniques

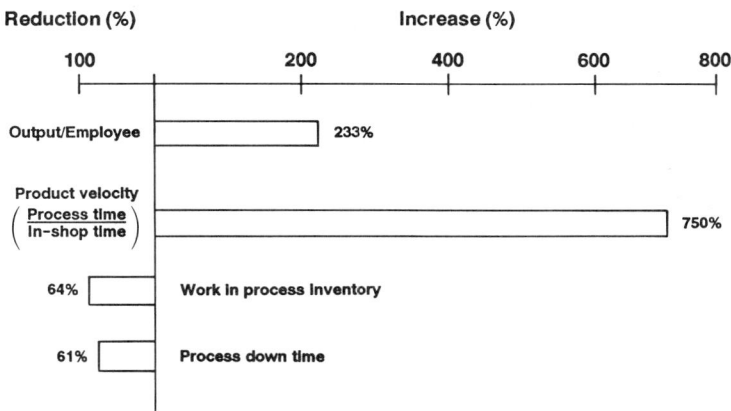

FIGURE 11   Denver Works productivity improvement initiative.

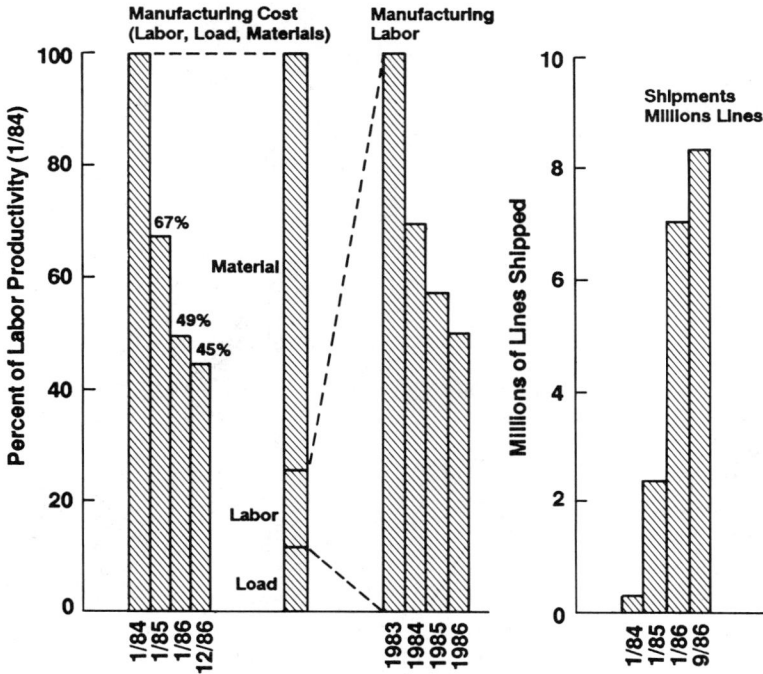

FIGURE 12 Oklahoma City Works productivity improvements, 5ESS™ systems.

product family is the largest factory-within-a-factory at Oklahoma City. The program followed the model shown in Figure 3. Information automation in the form of the PRISM systems was deployed in Oklahoma City beginning in 1984. The results are shown in Figures 12 through 14 for costs and capacity, cost of quality, and intervals. It should be noted that cost improvements were realized at about the same rate across

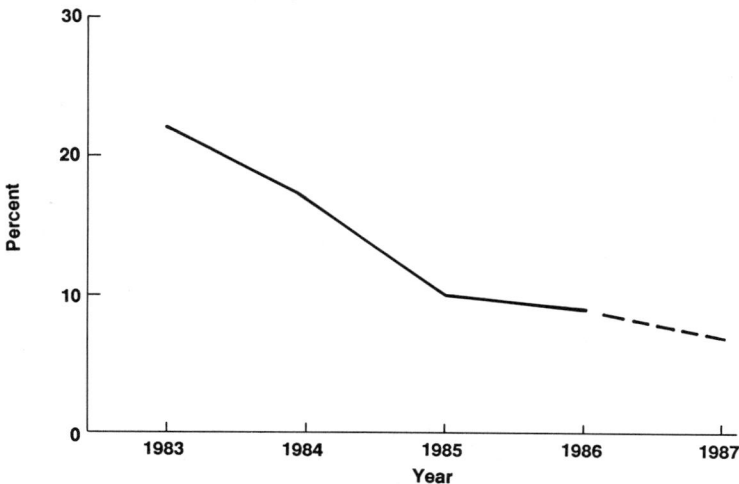

FIGURE 13 Oklahoma City Works quality/productivity (cost of quality includes prevention, appraisal, and failure).

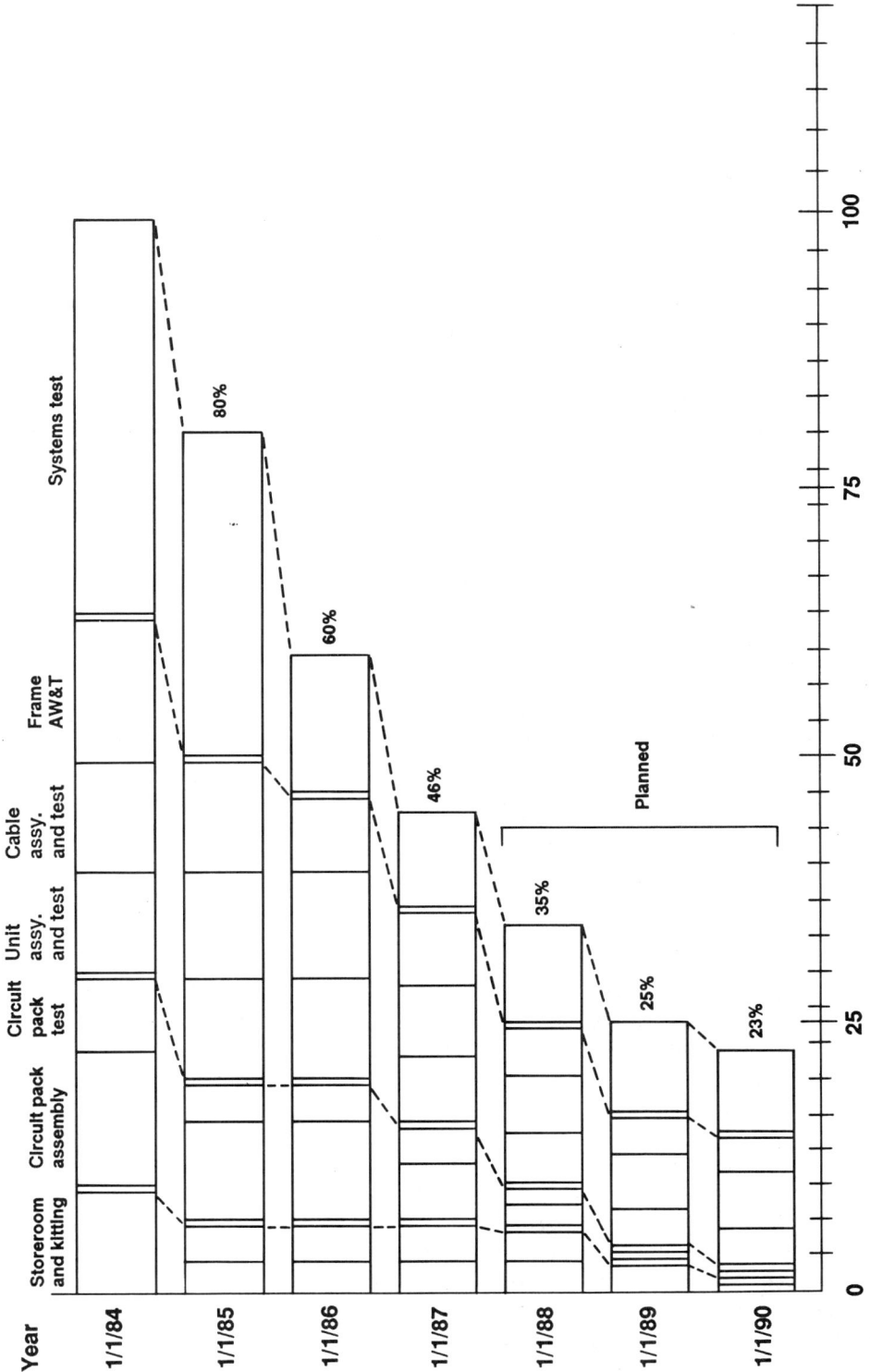

FIGURE 14 Oklahoma City Works Manufacturing intervals by shop, 5ESS™ systems. (Does not represent actual manufacturing intervals since a number of operations are performed in parallel.)

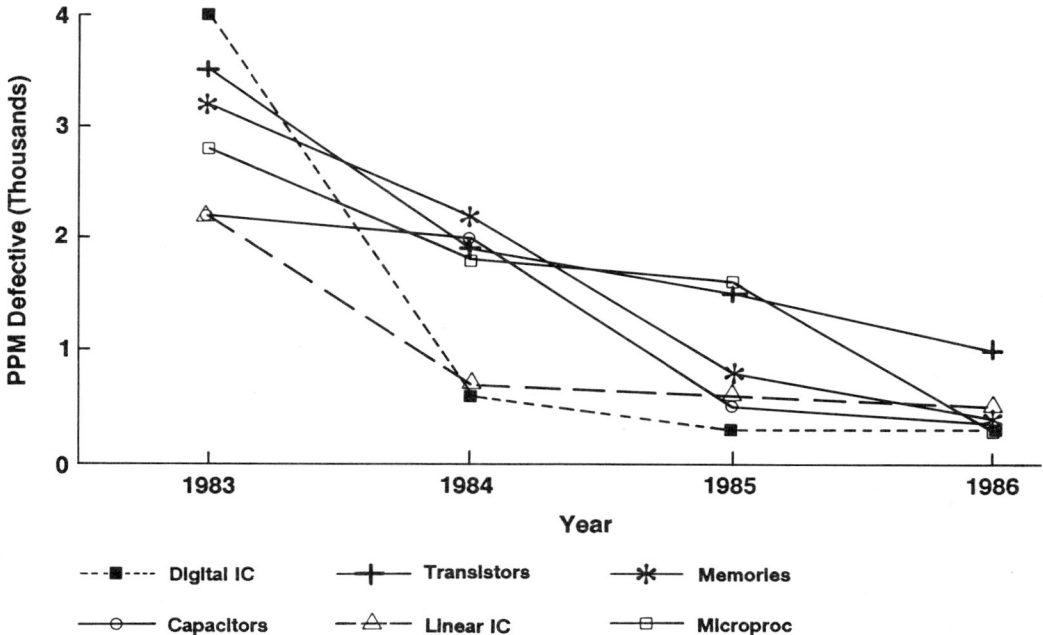

FIGURE 15   Component incoming quality at Oklahoma City Works.

labor, load, and material. Labor costs were halved, although no effort was made to directly replace human labor with robotics or any other physical automation. The labor reductions came from yield, quality, and throughput improvements. Material cost reductions were achieved by cooperative efforts of product designers and manufacturing engineers, including a major effort with all parts suppliers, both in-house and outside, to improve incoming quality levels. Figure 15 shows actual incoming component quality levels. This is exceedingly important, since 80 percent of the defects found during testing of the circuit packs result from problems in the quality of the components.

The knowledge derived from component engineering, process analysis, and FMA activities is now being used in the most productive of all activities, DFM. Figure 16 shows the value of DFM in yields of first production runs for new circuit packs. We consider these results for complex circuit

packs, using state-of-the-art devices, to be remarkable.

A study was conducted in 1986 to try to evaluate the elements needed for world-class manufacturing productivity. Table 2 ranks those elements based on available quantifications of savings and also by a survey of involved individuals. The latter included management, engineers, factory workers, and product designers. The fact that "teamwork" ranked highest in importance emphasizes the value of treating the human element in the process of improving manufacturing productivity.

The productivity improvement techniques described have also been applied to the manufacture of integrated circuits. As technological advances drive more and more circuit functions onto ICs and IC packages, the performance of the semiconductor fabrication processes will dominate manufacturing viability. These processes are more complex, yields are usually low, facilities are capital-intensive, and use of

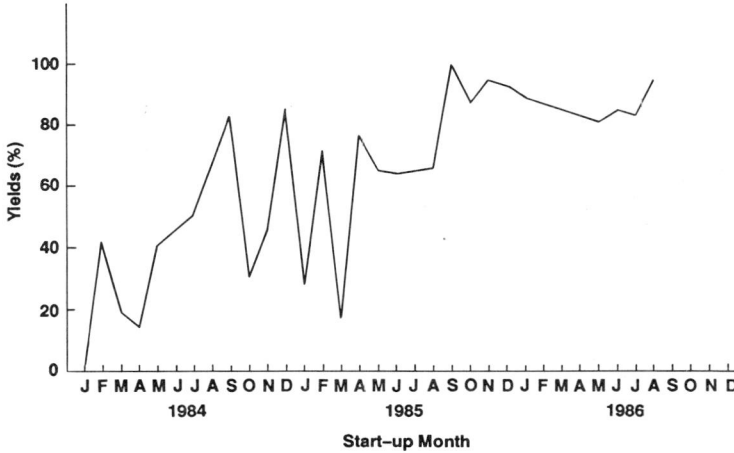

FIGURE 16   First production runs, new circuit packs.

state-of-the-art technology is normal. The manufacturing productivity realization model (Figure 3) is still valid, however. Figure 17 shows the results achieved in a silicon wafer fabrication process using this technique.

CONCLUSION

These successes are an "existence proof" of the value of first applying systems engineering disciplines to manufacturing projects, followed by information mechanization (automation plus integration), and

lastly the deployment of optional physical automation. There is significant room for improvement in manufacturing in the United States, and it is achievable within the capabilities of existing engineering disciplines and tools.

NEEDS AND OPPORTUNITIES

Many opportunities exist to further productivity gains in U.S. manufacturing. The following is a list of the most important areas for improvement:

TABLE 2   Elements of World-Class Manufacturing, 1986 Study of Oklahoma City Works

Ranking by Quantifiable Savings	Ranking by Consensus of Participants
1. Design for manufacture	1. Teamwork
2. Productivity improvement program	2. Quality management
3. Quality management	3. Customer satisfaction
4. Application of new process technology	4. Manufacturing systems engineering
5. Manufacturing research planning	5. Design for manufacture
6. Manufacturing systems engineering	6. Training
	7. Productivity improvement program
	8. Manufacturing resource planning
	9. Application of new process technology
Not measurable: Customer satisfaction Teamwork Training	

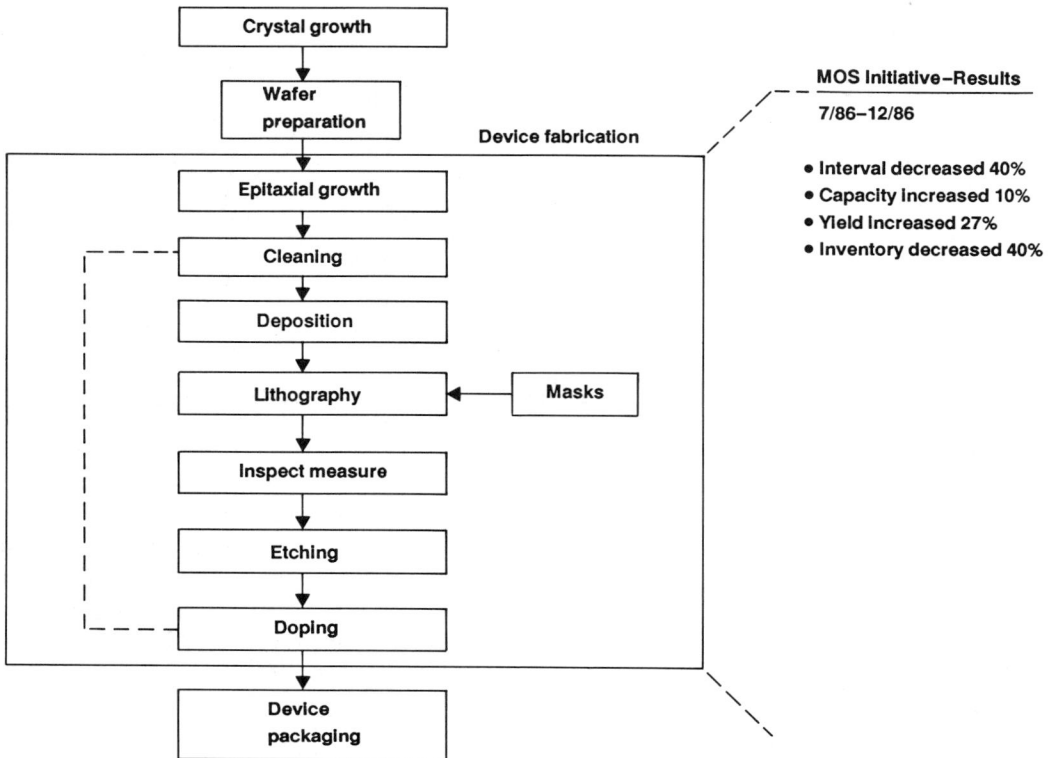

FIGURE 17   Performance improvements in semiconductor manufacturing processes.

• The systems engineering analytical discipline and its software tools (simulation, queueing, etc.), although readily available and rich in capability, are not as widely used as they could be. Probable causes for this are inadequate training of engineers and the lack of awareness of the benefits by corporate and manufacturing management. Efforts must be expanded to make the manufacturing community aware of these capabilities and to encourage the use of these systems.

• Design for manufacturability currently uses rather basic techniques. More advanced analytical techniques and tools for product designers are needed for linking product performance to process capabilities.

• Lack of computer interface standards among suppliers of production equipment impedes project implementation. Too much specialized "translation" software is required for CIM connection, thus reducing information processing effectiveness and adding to engineering costs. Semiconductor Equipment Communications Standard (SECS) and GM-S MAP standards activities are examples of attempts to rectify the situation.

• Opportunities exist for advancing the capabilities of production processes by applying new real-time computing technology capabilities to the following: process control; embedding software tools into CIM platforms, using system performance analysis, with "knowledge-based" and artificial intelligence techniques; and including advanced routing, scheduling, and comput-

erized *kanban* capabilities in CIM platforms to further the production flexibility.
Application of anthropomorphic robots for
flexible assembly may well serve the assembly of large items. Flexible approaches for
suprahuman machinery capabilities, however, are needed for electronics and pho-
tonics equipment and integrated circuit
fabrication.

REFERENCES

Eckel, E. J. 1986. Quality in AT&T Network Systems.
AT&T Technology Journal 65(2):30–38. Morris-
town, N.J.: AT&T Network Systems Group.
Jones, V. C. 1988. MAP/TOP Networking: A Foun-
dation for Computer-Integrated Manufacturing.
New York: McGraw-Hill.
Mayo, E. 1945. The Social Problems of an Industrial
Civilization. Cambridge, Mass.: Harvard Univer-
sity Press.

APPENDIX

AT&T MANUFACTURING SYSTEMS ENGINEERING TOOLS

ENGINEERING
Capacity

SESAME SCHEDULING EVALUATION SYSTEM FOR AUTOMATED MANUFACTURING ENVIRONMENTS
Determines the manufacturing interval, maximum utilization rate, number of shifts required, and number of machines required to maintain production rate for insertion machines that produce high-runner codes.

DCM DETERMINISTIC CAPACITY MODEL
Estimates the production capacity and machine utilization in a clean room as a function of the product mix.

LB/L LINE BALANCE/LAYOUT
Determines the minimum number of workstations and balances the workload among the stations to meet throughput requirements in assembly lines.

QNA QUEUEING NETWORK ANALYZER
Provides estimate of work in progress, interval, and utilization in general manufacturing systems using analytic queueing techniques.

Simulation Models

GSM GENERALIZED SIMULATION MODEL
Provides a generalized simulation model for complex manufacturing systems and estimates work in progress, interval, and level of utilization. This is a tool for addressing complex issues.

PAW PERFORMANCE ANALYSIS WORKSTATION
Builds graphics-based simulation models of queueing networks. This tool is used to answer design and engineering questions quickly.

Design

QCAP   QUALITY AND COST ANALYSIS PLAN

Analyzes the design and manufacturing process of an assembled electronic product. This tool is used to evaluate design, improve process, enhance quality, and decrease manufacturing cost.

OPERATIONS

Data Analysis and Monitoring

BCR   BAR CODE READER

Provides a systematic mechanism for tracking and monitoring portable, rechargeable bar code readers.

RTAS   REAL-TIME ANALYSIS SYSTEM

Uses statistical techniques to monitor and analyze a manufacturing process in real time and improve efficiency and output by identifying problems as they occur.

Quality

AGDAT   ADVANCED GRAPHIC DATA ANALYSIS TOOL

Displays and analyzes defect data using control charts.

PCAP   PROCESS CHARACTERIZATION ANALYSIS PACKAGE

Analyzes and summarizes large quantities of process and test data used by engineering.

ILIAD   I LEARN. I AID DIAGNOSIS

Employs an improved technique to "learn" effective troubleshooting methods *from* operators. Newly acquired repair knowledge is automatically added to ILIAD's knowledge base on a daily basis—ensuring its responsiveness to ongoing changes in the manufacturing process. In addition, ILIAD produces a number of analysis reports that help engineers and operators identify the root causes of many problems in the product as well as in the process.

Reliability

STAR   STATISTICAL ANALYSIS RELIABILITY TOOL

Identifies reliability problems and assists engineers with solving them by using statistical analysis techniques.

SUPER   SYSTEM USED FOR PREDICTION AND EVALUATION OF RELIABILITY

Predicts system reliability at any phase of product design using a variety of reliability modeling techniques.

Scheduling

DMIS   DYNAMIC MANAGEMENT INFORMATION SYSTEM
Estimates anticipated flow of product in clean rooms as an aid for scheduling of machines.

STBGT   SCHEDULING TOOL BASED ON GROUP TECHNOLOGY
Used to schedule lines with medium- to low-volume codes by characterizing product families.

LITES   LIGHTWAVE INTEGRATED TECHNIQUE FOR ENGINEERING SPANS
Blends fibers with differing performance specifications to meet customer specifications.

CAPS   CABLE PRODUCTION SCHEDULER
Sequences jobs through a production line choosing appropriately from candidate machines and produces machine assignments and job sequences for each machine.

FAS   FINAL ASSEMBLY SCHEDULER
Sequences final assembly in an attempt to smooth the demand on feeder shops.

LBS   LOADING, BUFFER SIZE, SEQUENCING
Quickly estimates production rate and machine use on circuit pack assembly lines and estimates effect of loading, lot sizes, and buffers on those rates.

NCMS   NUMERICAL CONTROL MACHINE SEQUENCER
Used to reduce processing times by making drilling, component insertion, and metal-punching operations more efficient.

FLEXIBLE MACHINING IN AN INTEGRATED SYSTEM

Arthur J. Roch, Jr.

ABSTRACT   Flexible manufacturing offers productivity, affordability, and enhanced quality. Yet the far-reaching benefits are by no means automatic. Massive modernization efforts can yield little net result in productivity improvement and cost reduction without adequate consideration of the specific application. Integrated systems offer the greatest potential for productivity improvement, yet this benefit must be designed into flexible manufacturing system applications to ensure success. New methods must be assessed in terms of production needs. Technology can then be designed into the system in response to the potential costs and benefits of implementation. Integrating and implementing technologies in new flexible systems to meet program needs is proving effective throughout the aerospace industry. This paper describes an approach to the successful application of flexible manufacturing system capabilities.

INTRODUCTION

American industry stands at the gateway of a new era in manufacturing—an era that began with the advent of hard automation. Transfer lines that were employed to reduce the labor hours consumed in making a product are being replaced by systems with greater flexibility. Flexible automation offers improved productivity and product affordability, yet it retains the benefit of improved quality that comes with hard automation. Not only does flexible automation offer increased process speeds, it also permits users to adapt rapidly to changes in product or product mix. Furthermore, flexible automation can compensate for a decline in the number of skilled manufacturing professionals that are available for employment in American factories.

With foreign competition an ever-increasing threat to American industry, flexible manufacturing is becoming a critical element in the American approach to achieving competitiveness in manufacturing. Automation technology is moving out of the laboratory and into the workplace just when American industry needs it most.

Flexible automation is particularly attractive to the aerospace industry. The aerospace marketplace is characterized by complex manufacturing requirements and small lot sizes. Capital investments that make sense in some industries—for example, automobile, farm equipment, and household appliance—must be reexamined when a lot size is 100, as in aerospace. A well-designed, automated flexible manufacturing system not only promotes economical production of small lot sizes but also eases the transition from one product to another.

The flexible machining system is the most prevalent form of flexible automation in use today for a number of reasons:

• Detailed part fabrication is easier to automate than assembly operations.
• Machining is perhaps the most mature detailed part fabrication technique.
• The American machine tool industry has enthusiastically pursued the advent of such systems to make its equipment more attractive to potential customers.

This paper addresses the broad subject of factory automation within the context of the aerospace industry. Machining systems are used as a practical example. The discussion concludes with the identification of critical emerging technologies for the further application of flexible machining systems.

Two such systems are discussed in this paper. The first is the Flexible Machining Cell (FMC) implemented by the LTV Aircraft Products Group (LTVAPG) in 1984. The second is the Integrated Machining System (IMS) that is currently under development at LTVAPG. This second-generation system is being developed using experience gained in establishing the first-generation FMC. The two systems provide substantial insight into the contents of a generic factory automation life cycle. The following sections describe this life cycle and relate the FMC and IMS to it.

PLANNING THE SYSTEM

Much has been written about the need for an overall implementation plan for the factory of the future. This paper assumes that those who would implement future factory automation systems will employ an architecture similar to that set forth by the Integrated Computer-Aided Manufacturing (ICAM) Project Priority 1105 (Air Force Wright Aeronautical Laboratories, 1984) and that they would embrace the philosophy of computer-integrated manufacturing (CIM).

The Factory Automation Life Cycle

The soundest foundation for implementation of a flexible machining system is a clearly identified and supported program need. Only with an unambiguous description of what is to be produced can the proper questions be identified and the technological answers determined. The life-cycle chain of events leading to a successful factory automation project involves a step-by-step process of design, analysis, and decision-making, with integral links between product and program requirements, return on investment (ROI) analyses, and design development, as indicated in Figure 1. Successful development can be seen as the logical and progressive result of a series of carefully executed tasks—each of which must be examined from both technological and financial perspectives.

Candidate System

The first challenge of implementation is identification of candidate systems. The ideal candidate factory automation system arises from thorough systems engineering analyses performed against specific program needs. The resulting concept is an accumulation of system design responses that have been refined and validated through simulation and modeling routines.

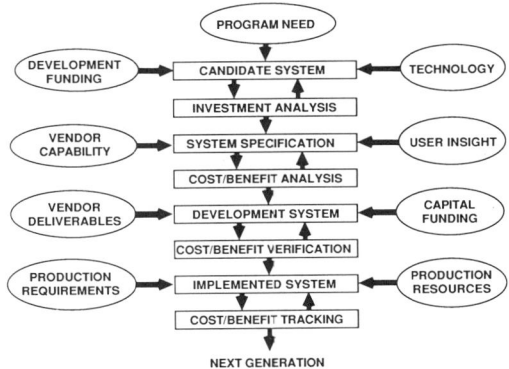

FIGURE 1   Factory automation life cycle.

System Specification

Financially verified system design elements are subject to further study during the development of specifications; the validity of this assessment is dependent on a strong knowledge base of vendor capability and user insight. Major system elements are defined (and refined) through a series of documented development tasks.

System specification begins by establishing a program office and generating a program plan. User support teams are then organized to produce a System Requirements Document (SRD) and a Request for Proposal (RFP) to be released to prospective vendors. The RFP includes formal specifications for the desired system and its major elements, such as computer control, process equipment, and material handling. Vendors are selected on the basis of program-driven evaluation criteria mapped onto an evaluation matrix. Once a vendor is selected, specifications are reviewed and revised, as necessary, according to the selected vendor system. Concurrently, the integrated system elements are verified through simulation and modeling.

The SLAM simulation language (Pritsker and Pegden, 1979) was used for this purpose, with user-written logic added to mimic the proposed scheduling logic. The primary uses of the model were to compute equipment and tooling capacity requirements, to develop and validate scheduling policy, and to serve as a focal point for documenting and resolving operating issues during the design phase.

In general, the methods and issue resolution role of simulation are often overlooked relative to the more quantitative type of output. The accuracy of the quantitative results—e.g., utilization, span times—is, of course, sensitive to the quality of the input data. This presents some problems when new technology is involved and when the methods and time data are poorly defined. A good reference on the use of simulation

in practice is Pope's discussion on the general role of simulation in aerospace manufacturing (Pope, 1986).

Cost/Benefit Analysis

While system specifications are being developed, a thorough cost/benefit analysis is performed to reaffirm the bottom line of implementation—maintaining an acceptable ROI. This analysis narrows estimates for the standard-hour content of expected part loads as well as nonrecurring costs.

Based on vendor proposals, the ROI is computed on the basis of an accepted model, and a capital expenditure is sought and approved. In approving the expenditure, management must consider the potential factory automation initiative from both technical and financial perspectives. The technology employed must be packaged into a sound business proposition.

The use of a cost center approach has simplified the task of properly accounting for costs associated with flexible machining systems. By creating discrete cost centers rather than relying on manufacturing pools, the flexible machining system is in a position to capture all costs associated with its operation. This accumulation of charges negates the dependence on overhead allocations based on diminishing amounts of direct labor. Although this practice satisfies the need for actual cost collection, it is not directly comparable with costs that would be expected for conventional machining systems, which have relied on pool rates and overhead allocations. This constitutes one of the basic difficulties in conducting financial analysis of flexible machining system installations. The creation of a synthetic present-technology cost center with conventional machining will allow a direct comparison without reliance on pool rates that may be either over- or under-inflated in relation to rates for the actual equipment replaced.

The basic measure of the flexibility of a cell can be quantified by an examination of how many discrete part numbers can be produced in the cell. The larger the number of different part numbers produced, the greater the flexibility of the cell. The cost-effectiveness of this flexibility can be directly related to the amount of resources necessary for the described part population. If the flexibility of a cell is limited to only 6 discrete parts from a population of 18, there is a need for 3 times the resources, in terms of capital, maintenance, direct head count, etc., than that needed for a cell that has the capability of handling the full population. Likewise, when compared with conventional stand-alone machining, the highly versatile flexible machining system uses less of these same resources to meet the machining needs of a given part population. These resources all have a cost that is both quantifiable and measurable.

Development System

The next step in the factory automation life cycle is the selection of a vendor and the commitment of capital resources. This action formally initiates the development process and gives the proposed system an upper management, and even corporate, visibility. The transition from business-as-usual to state-of-the-art must be carefully orchestrated.

At this stage of development, many technical disciplines from the purchasing company and from the system supplier(s) must work in tandem to prepare facilities and establish interfaces to other operations. Users receive training, start-up resources are arranged, and numerical control programs are built and proofed.

Cost/Benefit Verification

Even as vendors and functional departments become absorbed in converting the system into a reality, ROI remains the primary test for each decision. As the development system begins to take shape, contin-

uing cost/benefit verification is performed. Detailed estimates of standard-hour contents and actual vendor costs are computed, and nonrecurring costs are refined. Standard-hour content for new systems is derived from parametric estimates of existing machining standards adjusted for anticipated productivity improvements. This development constitutes the standard-hour content for a given part population that will be run in the new system. Adjustments or revisions are made as experience is garnered from actual system performance. This concurrent analysis, even at this early stage, allows design and operational changes to be made to ensure that the capital expenditures will be justified by the intended productivity improvement.

Implementing the System

A program team, composed of technology specialists and users with internal and external resources at their disposal, must be made responsible for implementing the system. This operational milestone, the creation of the program team, is signaled by a formal sign-over of the system's operation and control from the developing and implementing technology specialists to the organization(s) that will be responsible for operating the system.

Technical support is ongoing as the program team provides analysis and system debugging. Corrections and enhancements to software and hardware design, as well as facilities, are identified and coordinated by the program team. Modernization specialists continue in a consulting capacity to help the user organizations achieve and maintain productivity targets.

Cost/Benefit Tracking

Once the system is implemented, cost/benefit tracking can be based on real standard-hour performance along with actual recurring and nonrecurring costs. These data, tracked and maintained for the life of the system, provide the means to verify and sustain the ultimate factory automation life cycle goal of an acceptable ROI.

OPERATIONAL EXPERIENCE WITH A FLEXIBLE MACHINING CELL

The validity of the generic factory automation life cycle is demonstrated daily in LTVAPG's Flexible Machining Cell developed using the techniques described. The FMC became operational on July 2, 1984. The cell currently machines 568 different B-1B aft and aft-intermediate fuselage parts, with an economic order quantity of one. FMC has demonstrated a 3-to-1 productivity improvement ratio over conventional machining. Figure 2 shows the cell layout and current factory-floor view of this 40,000-square-foot facility.

As a first-generation flexible manufacturing system, the FMC provides state-of-the-art capabilities for the manufacture of detailed machined parts. The cell provides

• Total computer control by means of a centralized system
• Automated machining (drilling, boring, tapping, milling, and profiling) in one setup
• Automated pallet transport and machine loading and unloading
• Automated in-line part cleaning and inspection
• Centralized coolant and chip segregation and removal
• Computerized cell loading, scheduling, simulation, and cutting-tool control
• Palletized part loading and unloading at automated pickup and delivery stations
• Centralized electronic cutting-tool gauging and setup
• Cutter diameter compensation for numerical control programs
• Optimized part-family and tooling relationships
• An exceptional part mix and volume in an essentially unmanned environment

Implemented to meet specific detailed machined part needs for the B-1B program, the cell consists of

• Automated work changer carousels
• Machining centers
• Chip and coolant system
• Material handling system
• Cleaning module
• Inspection modules
• Automated storage and retrieval system for cutting tools
• Computer control system

Two 10-station carousels, obtained from Cincinnati Milacron, are employed to queue work in the load-unload area. In addition, each of the carousels has two load-unload stations. These stations are staffed by operators who receive their instructions from the computer control system through a CRT display.

Cincinnati Milacron also provided the eight unmanned computer numerically controlled (CNC), single-spindle machining centers contained in the FMC. The centers are capable of machining both aluminum and nonaluminum materials in four axes. Each center is equipped with an automatic tool changer with 90 cutter storage positions. The prismatic work area has maximum dimensions of 32 by 32 by 36 inches and a pallet load weight capacity of 5,000 pounds. Each machining center offers the capacity for three-axis and four-axis simultaneous contouring in the X, Y, Z, and B axes. In addition, the centers provide torque-controlled machining at variable spindle speeds ranging from 40 to 5,000 rpm. The centers are designed so that the actual machining operation takes place in an enclosed area, and thus chips and coolant are more easily collected and present less of an adverse impact on the factory environment.

A five-station automatic pallet shuttle system is built into each machining center to allow almost continuous machining operations. Spindles can be equipped with

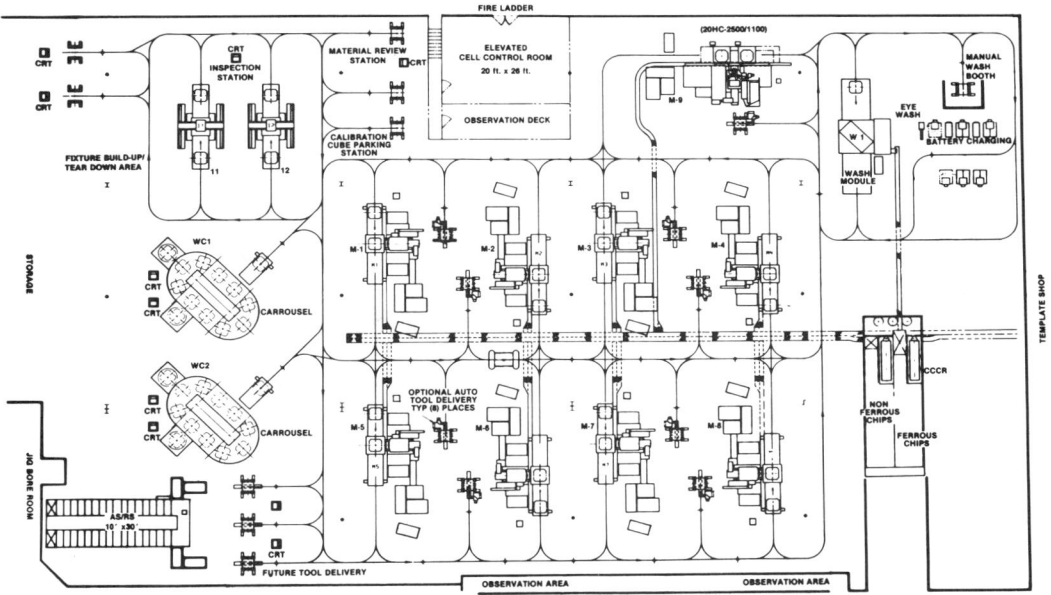

FIGURE 2   Flexible machining cell layout: (top) schematic, (bottom) factory-floor view.

part surface sensing probes to verify blanks with loaded CNC programs and establish first-cut positioning. The centers are also capable of automatically detecting broken tools and thus contributing to a reduction in scrap and rework.

A central chip and coolant system, designed by Henry Filters, Inc., is used for chip and coolant removal and collection from all machining centers and the cleaning module within the cell. Chip collection is accomplished without human intervention through the use of a dual-flume system installed in the FMC floor. A dual-flume chip and coolant system allows for the separation of aluminum and nonaluminum chips.

Pallet handling within the FMC is provided by a system of battery-powered, computer-controlled, wire-guided transport carts. The self-propelled carts, supplied by Eaton-Kenway, are under the control of their own computer, which is directed by the FMC host computer. The carts transport the pallet loads to and from the carousel pickup and delivery stations as well as to and from the other cell modules. The Taylor and Gaskin, Inc., cleaning module houses an unmanned automated liquid-wash and air-dry operation for preinspection cleaning of machined parts.

The two inspection modules, designed by Digital Electronics Automation, include electromechanical automated coordinate measuring machines for part geometry verification. Both cleaning and inspection modules are directed by the FMC host computer by means of specialized controllers using distributed numerical control programs.

A cutter crib automated storage and retrieval system (AS/RS) is located within the cell, as is the site of electronic cutter gauging and setup. Each cutter is logged into the FMC, given a bar code label, and computer controlled from that point forward. The cutter's dimensions, location, use, and maintenance are all monitored and directed by the cell's computer control system

working in conjunction with the AS/RS. Comprehensive electronic control of cutting tools is combined with compensation in numerical control programs for cutter dimensional variances due to wear. Tool use is maximized, while part quality is enhanced and scrap is reduced.

The LTVAPG FMC produces more part numbers with less manual intervention than any other such manufacturing cell in the world. The software and its associated computer system are designed so that the cell operates without manual intervention except at the load-unload stations. Productivity gains afforded with FMC are the result of reduced human intervention, improved material handling, enhanced throughput, and maximized control. The computer control system is the focal point of the cell's success. The unique FMC software architecture represents a new philosophy in computer control design. The master machining schedule is created by LTVAPG's business host computer. The FMC's control computer, a DEC 11/44, receives a 20-day window of work orders on a daily basis from the business host. The downloaded work orders are assessed, selected, and scheduled into the FMC by the computer control system.

When the data are downloaded to the FMC computer, the workload is automatically assessed to maximize use of machining center resources. This optimization includes detailed consideration of

- Due date
- Priority
- Material availability
- Numerical control and inspection part program availability
- Cutting tool requirements
- Fixtures and pallets
- Machining time

From the downloaded work orders, the FMC computer system selects enough work orders for a 24-hour production run. The FMC software analyzes the workload to obtain a 24-hour schedule for the cell's operation. The scheduler optimizes the 24-hour period to meet the production schedule, minimize cutter tool changes, maximize machine use, and reserve a selection of parts with high machining times for fabrication on the unmanned third shift.

FMC production capabilities are applicable to all aerostructures programs requiring aluminum and steel machined detail parts sized within the cell's work envelope. Flexibility in scheduling and operations provides detail parts in any quantity without penalties for short lead time or small quantities.

The FMC was developed and implemented under a Category 1, Phase III B-1B Subcontractor Technology Modernization Program. Internal funding provided development and implementation costs, with indemnification arranged through Rockwell International and the U.S. Air Force. As a cornerstone of LTVAPG's Multiproduct Factory of the Future, the FMC offers unparalleled flexibility in part load and quantity, substantial production savings, superior part quality, and exceptional performance.

A SECOND-GENERATION FLEXIBLE MACHINING SYSTEM

The success of the FMC has spawned new opportunities as a new generation of aerostructures evolves. A rapid change in aerospace manufacturing materials from conventional to exotic metals demands that a second-generation system, the Integrated Machining System, be implemented.

This second-generation system, although promising significant alterations in shop-floor activity, will make its greatest impact above the shop floor. The move to automate above the shop floor, first with FMC's control computer and now in the IMS, required LTVAPG to structure an information and function architecture that embraces total factory control. This hierarchical schema, shown in Figure 3, interprets inte-

grated computer-aided manufacturing in terms of program requirements and production needs. Control is driven to the lowest possible manufacturing level, permitting modular implementation of future systems, maximizing resource distribution and tracking, and optimizing information flow.

Simultaneously with FMC implementation, LTVAPG began preliminary studies of advanced manufacturing technology applications to future large-part and exotic-metal machining requirements. The resulting IMS concept, as shown in Figure 4, focuses on extending FMC manufacturing expertise to medium- and large-profile machined parts. The fabrication requirements for future metals, as found in candidate IMS parts, demand the use of five-axis computer numerically controlled high-speed machining (HSM) for aluminum and high-throughput machining (HTM) for titanium.

The most notable benefit of applying HSM and HTM technologies is the reduction of detailed part cutting time that will be afforded by next-generation spindles. HSM is projected to achieve a 6-to-1 improvement ratio, and HTM will afford a 2-to-1 improvement ratio. Furthermore, the productivity advantage of the spindles will be complemented by automated upstream preparatory tasks and downstream inspection and cleanup tasks.

The Ingersoll Milling Machine Company is under contract for turnkey implementation of IMS, with production start-up scheduled for early 1989. The IMS will require a floor area of approximately 150,000 square feet.

The IMS will encompass the functions for machined detail part fabrication of superplastically formed parts, castings, forgings, and plate stock, from the receipt and storage of raw material through the final unloading and tagging of a finished detail part, ready for hand finishing, nondestructive inspection, or processing. The IMS will incorporate design of a hierarchical computer control system to integrate the machin-

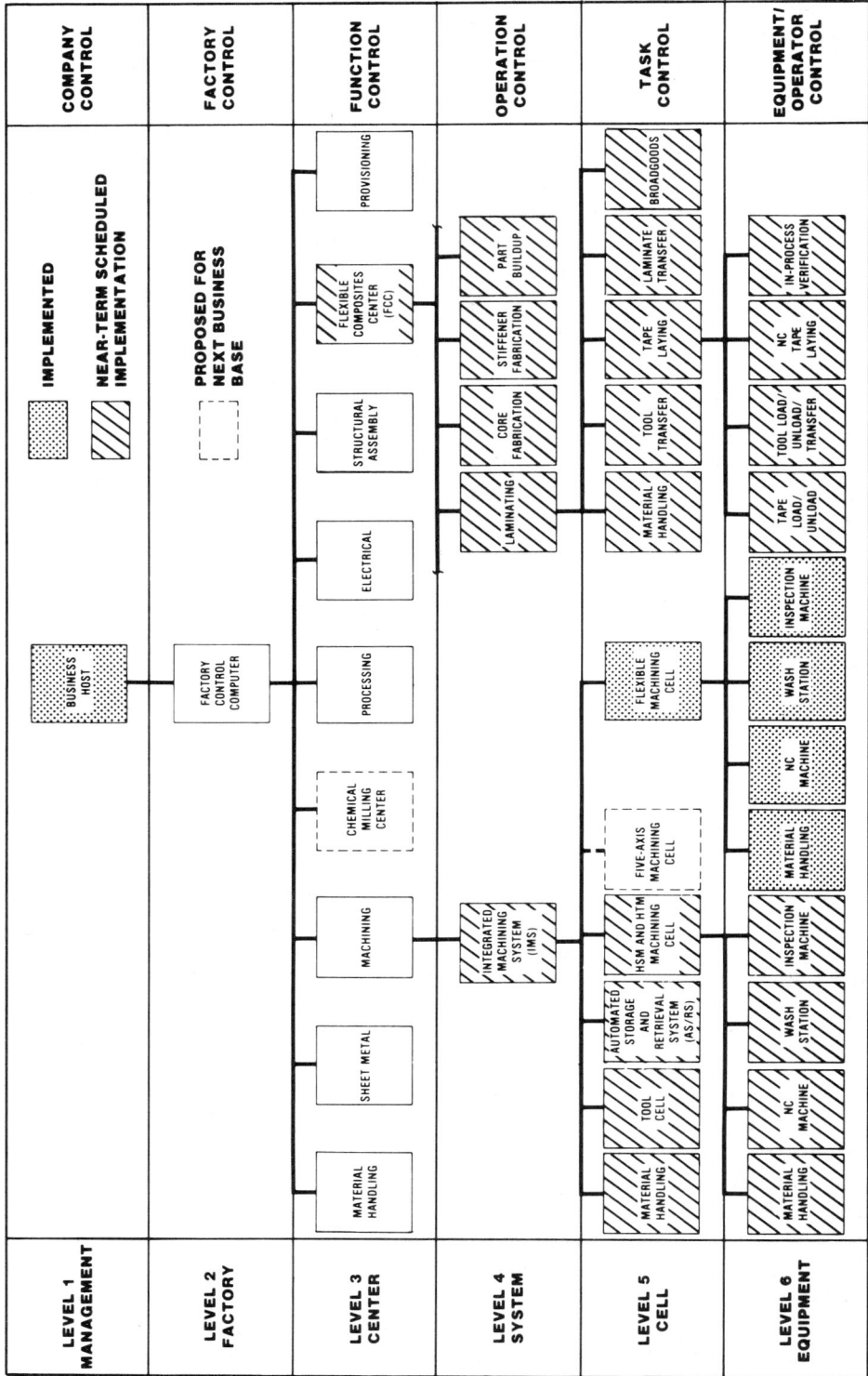

FIGURE 3 Factory-of-the-future hierarchical architecture.

ing functions with its support functions, as shown in Figure 3. In addition, it will control resources shared with LTVAPG's Flexible Machining Cell (FMC I). The following functions will be addressed in the IMS design for a second-generation Flexible Machining Cell (FMC II):

- Automated storage and retrieval systems for work-in-process (WIP), cutter components, and cutter assemblies
- Automated storage of pallets and fixtures
- Material handling systems for the transfer of WIP and cutters
- Cutter assembly buildup, preparation, and gauging
- Load-unload stations for pallets, fixtures, and parts
- Five-axis HSM of aluminum
- Five-axis HTM of titanium
- Automated washing
- Automated dimensional inspection
- Automated chip collection and chip transfer

The IMS project's primary design objectives are to apply leading-edge machining technology to

- Increase productivity of the LTVAPG machine shop in the fabrication of large, multiaxis aluminum and titanium parts
- Decrease WIP inventory of high-value material
- Reduce the direct cost per unit of output
- Improve product quality
- Improve throughput
- Promote timely delivery of parts to assembly

Productivity improvements resulting from IMS implementation will be the result of synergistic equipment and control operations along with advanced machining technology. Overall improvement estimates range as high as 6 to 1 over conventional methods.

FUTURE OPPORTUNITIES

Sufficient technology exists today to permit flexible machining system implementation. However, a number of emerging technologies will soon be available for incorporation into even more sophisticated flexible machining systems. These emerging technologies include

- Adaptive control
- Automatic cutter wear compensation
- Automatic broken tool detection
- In-process verification
- Robot-controlled deburring
- Automatic noncontact dimensional inspection
- Advanced cutting tools
- Generative process planning
- Computerized generation of as-manufactured configuration

Integration of these technologies with those existing today will result in a synergy that will further propel flexible machining systems into the mainstream of American manufacturing.

The design and application of such systems will continue to offer many challenges and uncertainties. To maintain focus on the task, the factory automation life cycle, as described in this paper, presents an abbreviated road map in the development process. Once one has started down the road to automation, it becomes difficult to identify stopping points. Focusing on productivity through repeated assessment of ROI provides the necessary signal. The system's ROI must define the level of automation. Design development in the factory automation life cycle must be driven by continuing cost/benefit assessment. The level of automation in the ultimate flexible machining system design is thus controlled through dynamic ROI assessment.

A successful system enhances profits and provides a competitive advantage. First-generation applications emphasized shop-

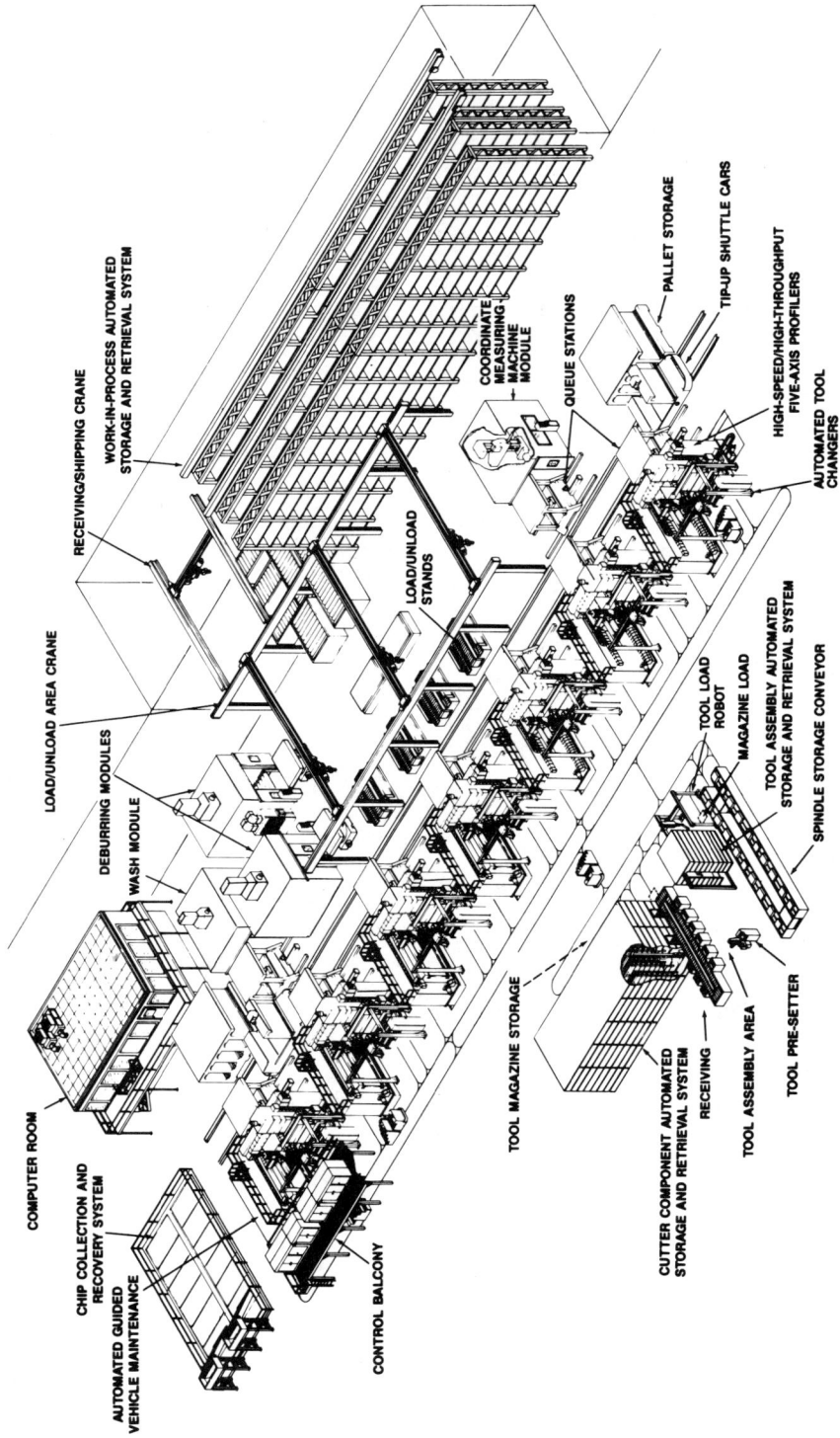

COMPUTER ROOM

CHIP COLLECTION AND
RECOVERY SYSTEM

AUTOMATED GUIDED
VEHICLE MAINTENANCE

CONTROL BALCONY

RECEIVING/SHIPPING CRANE

WORK-IN-PROCESS AUTOMATED
STORAGE AND RETRIEVAL SYSTEM

LOAD/UNLOAD AREA CRANE

DEBURRING MODULES

WASH MODULE

LOAD/UNLOAD
STANDS

COORDINATE
MEASURING
MACHINE
MODULE

QUEUE STATIONS

PALLET STORAGE

TIP-UP SHUTTLE CARS

HIGH-SPEED/HIGH-THROUGHPUT
FIVE-AXIS PROFILERS

AUTOMATED TOOL
CHANGERS

TOOL MAGAZINE STORAGE

TOOL LOAD
ROBOT

MAGAZINE LOAD

TOOL ASSEMBLY AUTOMATED
STORAGE AND RETRIEVAL SYSTEM

SPINDLE STORAGE CONVEYOR

CUTTER COMPONENT AUTOMATED
STORAGE AND RETRIEVAL SYSTEM

RECEIVING

TOOL ASSEMBLY AREA

TOOL PRE-SETTER

FIGURE 4   Integrated Machining System design concept.

44

floor automation and reduced production labor costs. These early efforts must be followed by second- and third-generation systems that focus on automation and integration of above-the-shop-floor activities, where a greater percentage of product costs is encountered; thus, modernization benefits are expanded.

Today's flexible machining systems applications are the forerunners of true computer-integrated manufacturing. CIM is essential to financial survival for American industry in the 1990s and beyond. In this new era of manufacturing, successful companies will be those American firms that developed long-range plans in the 1980s and pursued evolutionary implementation of those plans on the basis of specific business opportunities and technology applications.

REFERENCES

Air Force Wright Aeronautical Laboratories. 1984. ICAM Conceptual Design for CIM. Final Technical Report, AFWAL-TR-84-4020, Air Force Systems Command, Wright-Patterson Air Force Base, Ohio.

Pope, D. N. 1986. The role of simulation in aerospace manufacturing. Pages 2–97 in Proceedings, Vol. II, SME Ultratech Conference, Long Beach, California, September 1986. Dearborn, Mich.: Society of Manufacturing Engineers.

Pritsker, A. A. B., and C. Pegden. 1979. Introduction to Simulation and SLAM. New York: Wiley.

MATERIAL HANDLING IN INTEGRATED MANUFACTURING SYSTEMS

John A. White

ABSTRACT   A systems view of manufacturing is presented from a material handling perspective. A foundation for considering the next generation of material handling systems is established by first describing the role of material handling in manufacturing. The need for integrated systems is then considered. Subsequently, various aspects of designing, selling, specifying, and implementing integrated systems are treated. The subject of intelligent material handling is considered by focusing on its individual components: movement, storage, and control. Underlying this discussion is the assertion that the best material handling is *no* material handling. Finally, the status of the analytic techniques for treating material handling and the development needs for material handling are assessed. The paper concludes with the identification of a group of research and development tasks that can improve the analysis and design of integrated manufacturing systems.

INTRODUCTION

In most U.S. corporations, manufacturing is not accorded the high prestige given to product design or product engineering. And within the manufacturing community, material handling is not highly regarded. The best and brightest people in industry are seldom given manufacturing assignments, and it is a rarity that one of these is assigned to material handling. The career path to the position of chief executive officer neither begins with nor passes through material handling assignments; in fact, it is unusual to find it passing through a manufacturing assignment. Although a vice president for manufacturing might exist in a U.S. corporation, it is quite unlikely that he or she will be a member of the board of directors. This can be contrasted with the frequent election of the vice president for marketing or finance to the board of directors.

Not only has material handling failed to achieve recognition within the industrial community, but also it is seldom recognized by the U.S. research community as a bona fide research field. Since material handling is treated as a critical link in the manufacturing chain in Europe and Japan, it is not surprising that technological leadership in the field is to be found there. Although the technology for the automated guided vehicle originated in this country, the major advances over the past decade have been made by European firms. Major innovations in the design of such items as lift trucks, conveyors, and automated storage and retrieval systems also have come from outside the United States.

Despite this long neglect of material handling, there is reason for optimism. Increased attention is being given to it, and attitudes toward it are changing. It is now recognized that improvements in material handling must occur if manufacturing capability is to improve. Improvements will be critical in reducing inventories, improving quality, reducing cycle times, increasing productivity, and lowering costs.

Another reason for a changing attitude regarding material handling is an improved understanding of its role in integrated manufacturing systems. Once defined simply as "the handling of material," a systems view of material handling leads to a definition of "using the right *method* to provide the right *amount* of the right *material* at the right *place*, at the right *time*, in the right *sequence*, in the right *position*, in the right *condition*, and at the right *cost*" (Tompkins and White, 1984, p. 166). Because a material handling system cuts across cost centers and departmental boundaries, it functions as an integrating agent for manufacturing. For this reason, many consider the material handling system to be *the* systems integrator.

In the discussion that follows, the "systems view of manufacturing" is presented by addressing the definition of an integrated system, the need for integrated systems, including their design, specification, and implementation, the various aspects of automation, and finally material handling. The movement, storage, and control of material are examined in conjunction with integrated manufacturing systems. The paper concludes with a critique of the analytic tools most commonly used in designing material handling systems and an identification of development needs for future-generation manufacturing systems.

WHAT ARE INTEGRATED SYSTEMS?

A key issue that must be addressed is the definition of an integrated system. The term has been used for decades, yet considerable confusion exists regarding what is and what is not an integrated system. At the heart of the issue is the meaning of a *system*, which is defined here as a set of objects or elements, with relationships between them or their attributes, organized in such a way as to achieve a predetermined objective as they interact within their environment.

The elements of a system can be connected loosely or tightly. Although they can operate independently, so long as collectively they achieve a predetermined objective, an integrated system is generally one that is tightly connected. Put another way, the individual elements in an integrated system are synchronized. The connections can involve physical linkages by means of hardware or information linkages by means of computers and humans.

The emergence of computer-integrated manufacturing (CIM) has increased the attention given to the use of integrated systems. Unfortunately, much of the promise of CIM, though great, has not been realized. Some firms are advertising they sell CIM; others promote themselves as the turnkey CIM suppliers. With each playing a different game and using different rules, the only thing they have in common is the use of the same label (CIM).

A term-by-term examination of CIM, in reverse order, is useful. First, CIM applies to *manufacturing*. In this context, manufacturing is defined sufficiently broadly to include design and operating systems, as well as production and distribution activities. All value-added and support functions for the manufacturing operation are included, as are both direct and indirect labor.

The scope of CIM can be described best by a two-dimensional matrix. In the vertical dimension it includes the functions of product, process, and schedule design. In the horizontal it includes production, assembly, material handling, packaging, purchasing, quality control, production control, inventory control, maintenance, and distribution. CIM includes both planning and execution.

Second, CIM depends on *integration*. The synergistic benefits of systems integration that result in 2 plus 2 being greater than 4 is one of the promises of CIM. However, integration must be defined very broadly; it includes the interface and coordination of functions, the linkages of physical components, and the information handoffs that occur, both vertically and horizontally, throughout the entire organization. The breadth of integration requires more than the integration of two or three machines or workstations or the integration of two or three departments. CIM, when implemented to its fullest, will incorporate the entire manufacturing enterprise, including multiple production plants, suppliers, and customers.

Third, CIM depends critically on the use of the *computer* to perform the integrating function. Although it might include the physical integration of hardware components, integration of the information subsystems is essential to CIM.

It is worth noting, perhaps, the attributes that are not required of a CIM system. The production processes can be labor-intensive in a CIM environment, so long as they are integrated by means of the computer. One can achieve CIM without automating production or material handling, and one can achieve integration without the use of the computer, as, for example, in the Toyota just-in-time (JIT) system that relies on *kanbans*.

Many observers believe it is necessary for computer integration to occur using a highly centralized software architecture. Computer integration can be achieved by networking microcomputers, and information linkages can be performed using a combination of people and computers.

THE BARRIERS TO CREATING INTEGRATED SYSTEMS

The need for integrated systems in manufacturing has long been recognized (White, 1982). However, few truly integrated systems have been installed. Instead, automation islands, information islands, and even organizational islands have been formed. Some of the more common reasons for this follow (White, 1986b).

First, it is not easy to design and implement integrated systems. A detailed understanding of systems requirements and interactions must exist; many details must be considered; nothing can be left to chance. System complexity increases exponentially with the number of operations.

Second, designing and implementing integrated systems is a radical departure from tradition. Since the early 1900s, the approach to designing a system has been to divide the system into its basic components—e.g., operation, inspection, transportation, storage, and delay—and then to analyze the system in terms of these components, to use a division of labor, to design hierarchical organizations, and to form organizations into cost centers. To create an integrated system design from these many elements is not only difficult but has often proved to be impossible.

Third, there are few apparent rewards in designing and implementing integrated systems. Managers are frequently responsible for managing individual organizational units "by the bottom line." Few, if any, incentives exist to use team or systems objectives. Despite claims to the contrary, individual rather than team performance is frequently rewarded.

Fourth, "insurance" is preferred, often in the form of inventories, additional space, redundant equipment, and excess personnel. Minimizing risks seems to be preferred to maximizing gains. Thus, high inventory costs are often preferred to costs of stopping production. In addition, many believe a highly integrated system has no cushion for error; they believe a finely tuned "integrated engine" will experience considerable downtime (White, 1986c).

Fifth, organizational barriers must be overcome. Organization charts tend to create boundary lines. Many individual managers are reluctant to attack the overall system when they are being evaluated on the performance of the individual segment.

Sixth, the concept of the integrated system is not well understood. It has different meanings for different people, depending on their background and experience. To a hardware supplier, manufacturing systems are integrated if they fit together physically; to a computer supplier, they are integrated if the information systems share a common data base and provide real-time control; to a design engineer, integrated systems represent the combination of product design, process design, and schedule design (White and Apple, 1985). Each group views system integration differently, and each believes that system integration has been achieved by satisfying their view. Each frequently tends to think of an integrated system as a pipeline, rather than as a network of pipes. Consequently, integrated systems are frequently viewed as tightly connected, inflexible, and risky.

Seventh, the achievement of an integrated system requires a champion. A strong leader with a strong commitment to an integrated approach is necessary.

Eighth, few success stories and numerous horror stories exist. As Skinner noted, "The new, computer-based 'total systems' approaches to production management offer the promise of new and valuable concepts and techniques, but these approaches have not overcome the tendency of top management to remove itself from manufacturing. Years of development of 'the factory of the future' have left us each year with the promise of a great new age in production management that lies just ahead. The promise never seems to be realized. Stories of computer-integrated manufacturing (CIM) and new automated equipment disasters are legion; these failures are always expensive, and in almost every case management has delegated the work to experts" (Skinner, 1985, p. 55). Despite the highly publicized systems that have been successfully implemented by some leading firms,

the belief continues to exist that systems integration is expensive, risky, and complex.

Ninth, resources are often limited. In addition to a scarcity of capital, a critical shortage of qualified people frequently exists. The systems integrator should possess technical, management, and financial skills and an understanding of operational requirements. Among the technical skills needed is competency in developing control systems, for it is the control aspect that tends to determine success versus failure.

Tenth, some individuals are threatened by the use of integrated systems. In many firms, direct labor represents less than 15 percent of manufacturing cost. Materials and indirect costs contribute about 50 percent and 35 percent, respectively. Because integrated systems are aimed at eliminating gaps in the manufacturing process, they will result in a reduction in the labor that is used to close those gaps. Indirect costs should be reduced through the integration.

The evidence, however, is overwhelmingly on the side of the potential advantages for integrated systems. The promise continues to be great, even though current experience has not always been good. Computer hardware and software have been developed that allow information to be provided accurately and in a timely fashion for decision making. As a result, management's span of control has been expanded. Finally, integrated systems provide a mechanism to reduce the "fat" in today's manufacturing and distribution systems and to increase competitiveness.

Aristotle is credited with the observation that the whole is greater than the sum of its parts. In the case of integrated manufacturing, this claim is frequently repeated but with little evidence to support it. The costs and benefits of the whole cannot be accounted for by summing the costs and benefits of the component parts of the system. An overall assessment of the benefits must be made if the financial justification is to be made convincingly.

DESIGNING INTEGRATED SYSTEMS

In designing integrated systems, a delicate balance must be maintained in terms of the degree of risk one is willing to take with respect to leading-edge technologies versus proven technologies. A trade-off is required between the possibility of obsolescence of yesterday's technology and the possibility that tomorrow's technology might not be available on time, might not function as required, and might cost more than predicted. A concern with the risk of the unknown has perhaps been responsible for the lack of unproven technology in most integrated systems.

A number of challenges face the designer of integrated systems. The system must be kept simple and "requirements-driven." The ultimate user must be involved in the design process. The various design functions must be integrated. The system must be designed to accommodate automation, especially when the automation may be introduced at a later time. Excess insurance must be avoided. Systems discipline must be achieved and a long-term focus must be maintained while remaining responsive to short-term needs. And finally, one must avoid the "not invented here" syndrome (White, 1986a).

SELLING INTEGRATED SYSTEMS

In selling integrated systems to corporate management, a number of lessons have been learned. Among them are these: Design the whole, sell the whole, and implement the parts; be sure you know what you are doing; do your homework; thoroughly analyze the "do-nothing" alternative; involve the user; analyze thoroughly; sell aggressively; be realistic and thorough in estimating costs and benefits; ensure accountability; and find a champion (White, 1985).

SPECIFYING INTEGRATED SYSTEMS

For many, a significant change is needed in the way integrated systems are procured.

When a firm purchases an industrial truck, a conveyor, or a palletizer, detailed design specifications are used routinely. They are often written at the component level. However, when an integrated system is purchased, greater emphasis must be placed in the specification on what, when, where, and why, rather than on how, who, and which. The objectives are seldom defined well enough to allow the use of detailed design specifications. As firms obtain experience in implementing integrated systems and as increased standardization occurs in data bases, protocols, and communication interfaces, detailed design specifications will become useful in procuring integrated systems. Until then, functional specifications will be the preferred procurement instrument.

IMPLEMENTING INTEGRATED SYSTEMS

Careful planning is required if an integrated system is to be successfully implemented. This includes installing and debugging the system, training employees, and auditing the system to ensure that it meets requirements and is used properly. Although it is important to perform top-down design, it is critical that implementation be bottom-up. Designing from the top will ensure that the system is compatible with, and supportive of, the firm's business objectives. Implementing from the bottom up ensures that individual components can operate independently. The top-down design ensures that the operations will be synchronized.

As with most major changes, a transition from a segmentalist approach to an integrated approach will most likely meet with opposition. Hence, care must be taken to protect against premature rejection of the system. It is often the case that an existing system must continue to function while an improved system is being designed and installed. A phased implementation plan is required to avoid interrupting production. It is also difficult to maintain a long-term focus and simultaneously avoid making mistakes in the short term. Yet, these constraints must be accommodated, since a "greenfield" system is seldom feasible.

AUTOMATION'S REPORT CARD

Automation, in the form of automated storage and retrieval systems, guided vehicles, and automatic identification, has been used extensively in production environments for more than 20 years. Despite a lengthy record of proven successes, "horror stories" about the implementation of automation continue to appear. Reports of schedule slippages of months and years are not uncommon—slippages coupled with the doubling and tripling of cost estimates. Claims and counterclaims abound as to whether automated or conventional systems are the best approach. Why is automation such a controversial topic? A number of observations may help answer this question (White, 1986a).

First, the most significant lesson learned is that the primary benefit of automation is the systems discipline it imposes. An automated storage and retrieval system (AS/RS) is a highly disciplined technology when compared with the conventional system. To the extent that the same level of discipline is obtained using more conventional approaches, it is difficult to justify automation. In fact, it is difficult for automation to compete against a well-managed, highly motivated work force. But it is also true that very few well-aged, highly motivated work forces exist in the United States.

Second, the technology of automation should not be singled out for blame. Rather, the design, implementation, and use of the overall system should be examined before making a judgment of culpability. It is an uncommon exception when the technology will not work. It is more usual that the scope of the application was too broad, the wrong technology was adopted, or the technology was being applied incorrectly.

Third, too many automated systems are designed "for show, not for dough." Too little attention is paid to the tangible benefits derived from capital investments in automation. There is a tendency to buy more hardware or software rather than to invest in a careful system design to ensure that the investment will pay dividends. As a result, automation is often overpromised and underdelivered (White, 1984).

Fourth, rather than being designed to meet a set of basic requirements, system designs are often determined by existent or, in rare cases, anticipated technology. They are solution-driven systems instead of requirements-driven systems.

Fifth, the real system requirements are seldom well defined. The supplier is often asked to do "free design" and quote a price for satisfying a user's needs, without those needs being specified quantitatively in terms of performance requirements.

Sixth, the automation bottleneck is often software, not hardware. The control system is often designed with an excess of control. Rather than err by undercontrolling the system, the designer tends to overcontrol. Instead of designing the control system on the basis of the maximum amount of information possible, it should be designed on the basis of the minimum amount of information required.

Seventh, a new automation system becomes the vehicle for solving tangential problems—for example, purchasing, receiving, shipping, accounts payable, accounts receivable, inventory control, and other functions. The result can be that a major software task is assigned to a firm whose area of expertise is represented by less than half the total effort. Suppliers are often expected to satisfy the users' gluttonous appetite for sophisticated software. "Software-smart" suppliers typically are not "material handling hardware-smart," and vice versa.

Eighth, the user often lacks the discipline to "freeze the design" following the procurement decision. As a result, requirements are often added while the system is being designed. Design changes are both expensive and disruptive to the delivery schedule. Also, it is often the case that changes made during the design and implementation process are "spur of the moment" responses to a rare event.

Ninth, the computer hardware to be used is frequently specified before the computing requirements are defined. As a result, it is frequently the case that the computer is undersized in the initial quotation and that the error is not discovered until the system is installed and fails to handle the requirements.

Tenth, unreasonable, unnecessary, and expensive requirements are placed on the system to "never fail!" Reliability, availability, and maintainability values are specified arbitrarily, rather than by considering the cost impact of purchasing redundant computers. The cost of having "full-system recovery" within x seconds of an unscheduled computer stoppage should be assessed before specifying the value for x.

INTELLIGENT MATERIAL HANDLING

Desirable attributes of a material handling system were previously given as the right *method, amount, material, place, time, sequence, position, condition,* and *cost.* A material handling system can also be defined as *moving, storing,* and *controlling* material. Despite recent improvements in material control that have transformed it from a mundane task to a sophisticated, state-of-the-art activity, one must not lose sight of the fact that the best material handling is no material handling. Although material movement, storage, and control are often taken for granted, an idealized manufacturing system would be one in which material did not have to be moved, stored, or controlled. Although the ideal of completely eliminating material handling may not be possible, it is certainly true that handling less is best. More specifically, less material movement is best, less material storage is best, and less material control is best.

Less Material Movement

In what sense is "less material movement best"? For some applications, "less" might mean less frequently, less material, less distance, fewer interruptions, less manual intervention, or less variability. The objective of continuous flow manufacturing is to transform discrete parts manufacturing into a nonstop process in which material moves steadily through the production process. In general, quality tends to decrease and costs tend to increase as material movement increases. For these reasons, moving material less *frequently* is best.

The successes of the Toyota Production System and JIT have resulted in the elimination of much of the material inventory. As a result of a renewed focus on the amount of material being moved, the size of the unit load and the number of moves performed have both decreased. Under these conditions, moving less *material* has proved best.

Another result of JIT is improved "pipeline" management. Suppliers are locating production facilities closer to their customers, and the distances between successive production operations are being reduced. The reduction of the amount of inventory in the "production pipeline" reduces overall costs and increases the ability of the production system to respond to changing requirements. It has become better understood that moving over less *distance* is best.

To eliminate waiting, the speed with which material moves must match the speeds of the processes it feeds. In the past, little attention has been given to the material waiting for processing. Instead, the focus was entirely on maintaining constant production. It was acceptable to have large buffers of material, if this was necessary to ensure that machines would not wait for material. Parts shortages and processing interruptions would not be tolerated, regardless of the cost. In the identification and elimination of sources of variation in production, the interface of the material handling system with the production processes has proved to be a fertile area for improvement. It has been shown that moving material with fewer *interruptions* is best.

Although U.S. manufacturers have available vast arsenals of sophisticated technologies, a high percentage of human activity in manufacturing is engaged in performing material handling. An analysis of the basic work elements performed by human operators reveals that most could be categorized as material handling. Simple pick-and-place activities continue to be performed by people. For many firms, the impact of manual material movement on quality, productivity, and cost is such that for them, moving less *manually* is best.

A major impediment to improved material handling is the lack of standardization. Standard containers, pallets, tote boxes, cases, and cartons do not exist. Except for bar code standards adopted by the Department of Defense, the automotive, health, pharmaceutical, and meat-packing industries, little standardization has occurred in the past 20 years. One example of a failure to standardize is the design of cases to hold size 303 cans, a standard can in the food industry. The Food Merchandisers Institute and the Grocery Manufacturers Association have identified 32 separate sizes of cases that are used to contain 24 of these cans (Sims, 1986).

Few material handling systems actually handle material. Instead, they move, store, and control containers (pallets) in (on) which material is placed. Hence, the basic building block of a material handling system is the container or pallet being handled. A failure to standardize on this unit creates a wide assortment of rack openings, conveyor widths, and lift truck sizes. Because the material handling system must be designed to move so many different containers, moving less *differently* is best. Standardization would produce dramatic reductions in material handling costs.

Less Material Storage

Just as intelligent material movement is *no* material movement, so intelligent material storage is *no* material storage. Although it may not be possible to eliminate material storage, an intelligent material storage system should be designed with the objective of "zero storage." As with moving material, "storing less is best." This can mean less frequently, less material, less distance, less volume, less money, or less routinely.

Storage is required when there are imbalances in the flow rates of successive operations. A balanced production system and a reduction in the number of parts, subassemblies, and assemblies that require storage will reduce the requirements for material storage. As an example, if several products are identical before a particular production operation, such as painting, then any in-process inventory storage should occur before the painting operation. Parts standardization is another way of reducing the number of parts requiring storage. Storing less *frequently* is best.

The emphasis on storing less material includes all types of material—e.g., raw material, subassemblies, in-process material, finished goods, tooling, supplies, equipment, and other support facilities. Storing less *material* is best.

Because of the need for improved response time in delivering materials to the point of use, a number of organizations are shifting from centralized storage to distributed storage. Past arguments that centralized storage reduced aggregate inventory levels are not necessarily valid in a real-time material control system; with real-time control, it is possible to manage distributed storage systems to the same degree that centralized storage systems were once managed. By using distributed storage, material is stored closer to the user. Hence, storing at less *distance* from (or closer to) the user is best.

Material storage involves more than just storing material. Space is required for clearances, aisles, storage racks, and containers. It is not uncommon in a well-designed pallet rack storage system for the material being stored to represent, on the average, less than 10 percent of the cubic volume of space consumed. As the cost of space increases, it becomes obvious that storing less *volume* is best.

The cost of storing material tends to be underestimated. Too many believe the only cost of material storage is the opportunity cost associated with the cost of the material itself. Thus, as noted previously, inventories have become pseudo-insurance policies. More discriminate management is needed; increased differentiation is needed between high-cost and low-cost materials to reduce inventory investment. By recognizing that materials are fiscal resources, it is evident that storing less material means storing less money, and storing less *money* is best.

In many cases, material storage has become a routine activity. In particular, the very existence of material storage has become a "given" when manufacturing systems are being designed. When major capital investments are made in automated storage systems, subsequent modifications of the production system are designed around the storage system; too often, the "sunk cost" invested in the storage system continues to influence future design decisions. Furthermore, the storage methods used are frequently routine; in many cases, manual storage methods are used, with subsequent inefficiencies in use of space, difficulty in maintaining accurate inventories, and excess time spent looking for material. As a result, it has been concluded that storing less *routinely* is best.

Less Material Control

Although many operators agree that the best material movement is *no* material

movement and the best material storage is *no* material storage, few agree that the best material control is *no* material control. The trend in designing material control systems is to overcontrol rather than to undercontrol. Since software has frequently become the bottleneck in implementing automated material handling systems, it is concluded that "controlling less is best." In this case, "less" means less centrally, less collectively, less complexly, or less frequently.

In a highly centralized control system, the production system becomes extremely vulnerable to start-up delays, software "bugs," human resistance, and information overload. In the extreme, the centralized control system is expected to be omniscient, immutable, omnipotent, and veracious. To the extent it is not, difficulties arise. Thus, controlling less *centrally* is best.

Several years ago, it was recommended that different things be moved differently, stored differently, and controlled differently. The objective of the recommendation was to recognize the inherent differences in movement, storage, and control requirements when designing a material handling system. The differences in the physical characteristics make it obvious that an item weighing 1 ounce should be moved differently than an object weighing 1 ton, and an ice cube must be stored differently than a diamond. Although many material handling systems do move and store different things differently, they seldom control different things differently. One can find examples of real-time control being implemented for an inventory consisting of more than 1 million part numbers when less than 100,000 of these experienced any activity over a 2-year period. The system should control material *selectively*, not *collectively*. Thus, many believe that controlling less *collectively* is best.

As noted earlier, the tendency is to overcontrol rather than to undercontrol. Similarly, there is a tendency in designing computer control systems to make them overly complex. Complexity can occur with either human or computer control. However, it tends to occur most easily when using computer control. Simplicity must be the hallmark of control systems design; controlling less *complexly* is best.

"The best material control is *no* material control" can mean that the best material control occurs when there is no need to control material. Eliminating the need for control simplifies the design of the material handling system. Control is needed to deal with variances; by identifying and eliminating the sources of variance, the need for control systems is eliminated. System discipline should be achieved without the use of computer or human control. Also, combinations of human control and computer control are preferable to computer control alone. If the frequency with which control needs arise can be reduced, then controlling less *frequently* is best.

MATERIAL HANDLING: ANALYSIS AND DEVELOPMENT

We now review the status of analytic tools for material handling and the development needs for future-generation material handling systems. Both design and operational issues are considered.

The Status of Material Handling Analysis

As noted previously, relatively little research has been performed on material handling problems. Although little attention has been given to a number of fundamental design and operating issues, considerable effort has been devoted to the development of analysis tools applicable to the design and operation of material handling systems.

Among the various analytic tools commonly used to facilitate the design of material handling systems, simulation is the most popular. Simulation is also being used to assist in the management of operating systems. Recent advances in simulation modeling have included animated output and the interface of simulation with computer-aided design (CAD). As a result, a model of a material handling system can be used to observe its simulated performance. The combination of color graphics, animation, and the CAD interface has transformed simulation into a powerful design, operating, and marketing tool.

Advances in the ability to solve large-scale simultaneous equations quickly have enhanced the feasibility of modeling analytically large-scale material handling systems. Recent work in queueing analysis has been particularly important in increasing computational capability. The relevant research also includes studies on closed queueing networks (Solberg, 1977; Suri, 1983), open queueing networks (Shanthikumar and Buzacott, 1981), the control of queues (Baras et al., 1985; Gonheim and Stidham, 1985), networks of queues (Whitt, 1983a, 1983b), and mean-value analysis (Suri and Hildebrant, 1984; Hoyme et al., 1986).

Mathematical programming formulations of material movement and storage systems have been developed. However, since most real-world problems result in nonlinear, integer programming formulations, these formulations are usually solved using heuristic approaches. The design of a closed-loop conveyor network linking multiple machine tools is an example of a situation that is difficult to model and solve exactly, since it requires that decisions be made on the buffer sizes at each machine, the number and locations of "by-passes" or "crossovers" on the network, and the control logic to use. A number of scheduling and sequencing decisions arise in operating a material handling system.

Unfortunately, little attention has been given to the operation of such systems. For example, little work has been performed on determining the "optimum" sequence of storages and retrievals for a rotary rack with robotic loading and unloading. The same is true for a microload automated storage and retrieval system. Even less attention has been given to determining "optimum" control strategies for material handling subsystem combinations. Beyond the explicit consideration of material handling problems, there has been little attention to material handling in the formulations of related problems. For example, few formulations of scheduling problems have as their objective the minimization of material movement and material storage. Unfortunately, there is little concern a priori for the impact of scheduling decisions on material handling.

Although inventory control has received considerable attention from researchers, the problem formulations have often failed to incorporate a number of significant aspects of the inventory problem as it occurs in manufacturing. For example, the determination of production lot sizes and unit load sizes, the use of centralized versus distributed storage, and the conditions under which kitting should be used have not been addressed from a systems perspective. Inventory formulations have tended to be too narrow in scope.

In general, too much academic research has been "solution-driven" rather than "problem-driven." So-called applied research has been performed on contrived problems rather than on real problems. Rather than becoming intimately familiar with the problem and defining a research agenda to address it, the initial focus has been on the solution methodology. As a result, the problem has been viewed from the perspective of the solution and formulated accordingly.

In terms of material handling technologies, little research has been devoted to the selection of the appropriate technology for specific applications. As an example, the use of asynchronous material handling equipment has become quite popular. Asynchronous alternatives include the use of automatic guided vehicles as assembly platforms and for performing general transport functions; "smart" monorails for moving parts between workstations; transporter conveyors for controlling and dispatching work to individual workstations; robots for performing machine loading, case packing, palletizing, assembly, and other material handling tasks; microload automated storage and retrieval machines; cart-on-track equipment for moving material between workstations; and manual carts for low-volume material movement activities (National Research Council, 1986).

Choices are being made by the automotive industry, for example, between the use of power-and-free conveyors and automated guided vehicles in assembling cars. Yet, no comprehensive models exist to support such decisions. The decisions tend to be based on intuitive appeal and anecdotal evidence. Electronics firms store components in carousels, in miniload automated storage and retrieval systems, in microload automated storage and retrieval systems, and in bin shelving. Although each of these has different performance characteristics, investment costs, and operating costs, little has been done to guide the material handling system designer in making selections among the alternative technologies and in configuring and operating each alternative optimally.

Experience has shown that it is difficult to develop economic models of technology selection decisions. Although it is easy to assess the economic impact of incremental changes in the design configuration, the sales price for each technology alternative is dynamic and is determined by prevailing market conditions, and the decisions are often based on consideration of multiple criteria. Hence, prescriptive models of technology selection decisions are, at best, aids to decision making.

Even though some attention has been given to interstation handling, little attention has been given to material handling at the workstation. As noted previously, a considerable amount of human activity in manufacturing is used to move, store, and control material. In many cases, the operator's vision is used to locate a part and determine its orientation in order to grasp and position it for the next operation. In all likelihood the orientation of the part was known at some previous point in the process. However, either the orientation changed or the information was not captured. Hence, the operator must regain physical control of the part. The cost of *regaining* versus *retaining* physical control of material has not been adequately addressed.

Although the need exists for automatic storage and retrieval of individual items, few equipment alternatives are available, and those that are available are not widely accepted. The technology void has existed for a number of years but does not appear to represent an area of current interest to material handling equipment suppliers.

Material Handling Development Needs

The assessment of the development needs for material handling is limited to the focus of this volume—namely, the design and analysis of integrated manufacturing systems. Hence, only those material handling technology gaps that affect the design of integrated manufacturing systems are identified.

The following list of development needs is organized into three general categories: material handling systems design needs; material handling interface needs; and material handling hardware and software needs.

Systems Design Needs

The following items relate to material handling systems design needs:

- Engineering workstations for designing material handling systems
- Expert systems for designing material handling subsystems
- Preprocessors that will create simulation programs from material handling systems designs

- Preprocessors that will create "optimum" control systems designs from simulation programs
- Increased understanding of the performance characteristics of material movement and material storage technologies
- Performance models for collections and combinations of material handling technologies
- A method of determining the ease or difficulty of moving, storing, and controlling a particular part or product
- Decision rules for retaining versus regaining the physical orientation of individual parts
- Decision support systems to assist the designer in determining the sizes and locations of material storage points and the unit load size to be moved between workstations
- Network generators for a variety of material movement alternatives, both synchronous and asynchronous

Interface Needs

The following items relate to the interface of material handling with product and process design, manufacturing systems, and shop-floor control:

- The incorporation of material handling considerations in the decision support systems used in product and process design
- The incorporation of material handling considerations in the formulations of manufacturing systems models
- Integration of distributed material handling control with shop-floor control systems
- Human supervisory control systems for distributed, automated material handling systems

Hardware and Software Needs

The following items address hardware and software needs:

- Automated material handling systems that recover automatically from significant disruptions

• Modular and flexible material handling equipment for use in moving and storing a variety of components and products
  • Direct identification technologies
  • Automated storage and retrieval systems for individual items
  • Path-free automated guided vehicles
  • Container and hardware interface standards

Within the first group of development needs, the targeted result is the development of an engineering workstation for use in designing material handling systems in manufacturing and distribution. To accomplish the objective, expert systems must be developed. The development of expert systems will be aided by the emergence of preprocessors to create simulation programs from CAD-generated material handling systems and to create the control system directly from the simulation program.

Before expert systems can be developed, there must be increased understanding of the performance characteristics of material movement and material handling technologies. To gain that understanding, performance models must be developed for logical collections and combinations of material handling technologies.

Among the decision support systems needed in designing material handling systems are those that provide assistance in determining unit load sizes, material storage locations, buffer sizes, material flow paths (networks), and positional control required for parts movement.

A major impediment to developing expert systems for designing material handling systems is the paucity of metrics for gauging the ease or difficulty of moving, storing, and controlling parts of various designs and configurations. "Design for handling" research similar to that performed in support of "design for assembly" is needed.

As noted previously, because of the serial nature of manufacturing systems design, it is generally the case that few degrees of freedom remain for the material handling systems designer after the product and process designers have completed their work. One solution to the problem is to provide the product and process designers with design tools incorporating material handling considerations. The second category of development needs addresses this issue.

In addition to the need to interface material handling design decisions with process and product design decisions, it is also important to incorporate material handling considerations in scheduling algorithms, shop-floor control systems, and inventory control models.

Finally, any consideration of material handling interfaces must include the human interface. In 1987 the human is still the predominant material handler in industry. However, that is not the interface issue addressed here. Instead, we are concerned with the human operating as a supervisor for the automated material handling system in a highly distributed environment.

Among the hardware and software development needs are the need for automatic recovery, automatic identification, automatic ranging and guidance, and automatic storage and retrieval of individual items. In addition, a critical need exists for the development of standards for containers and hardware in material handling.

A RECOMMENDED APPROACH

With the material handling development needs defined, what approach should be taken? The following five recommendations are given as research guidelines.

First, a problem-driven approach is recommended. Hence, a joint industry-university research team is needed. The team should consist of individuals who understand the problems, individuals who understand the methodologies, and those who can function as bridges between these two groups. The research should be conducted on factory floors and in laboratories rather than in offices and conference rooms.

Second, a cross-disciplinary approach is needed. The development needs listed will require the expertise of computer scientists, electrical engineers, industrial engineers, mechanical engineers, and systems engineers, among others.

Third, the research should be results-oriented. Specifically, a particular development should be targeted as the end-item deliverable from the research. Throughout the research, the targeted goal should be kept in mind.

Fourth, the first end-item deliverable should be the engineering workstation for designing material handling systems. The first ten development needs listed should provide a starting point for the research team. However, it is likely that additional needs will be identified in the conduct of the research.

Fifth, the research should be generic for discrete parts manufacturing. However, to facilitate the research it will be helpful to focus initially on a particular scenario. Experience indicates that the material handling problems encountered in industry are quite similar, despite the differences that exist in materials, tooling, processes, and levels of technology employed.

REFERENCES

Baras, J. S., A. J. Dorsey, and A. M. Makowski. 1985. Two competing queues with linear costs and geometric service requirements: The uc-rule is often optimal. Advances in Applied Probability 17:186.

Gonheim, H., and S. Stidham, Jr. 1985. Control of arrivals to two queues in series. European Journal of Operational Research 21:399.

Hoyme, K. P., S. C. Bruell, P. V. Afshari, and R. Y. Kain. 1986. A tree structured mean value analysis algorithm. ACM Transactions on Computer Systems 4(2):178.

National Research Council. 1986. Toward a New Era in U.S. Manufacturing: The Need for a National Vision. Manufacturing Studies Board. Washington, D.C.: National Academy Press.

Shanthikumar, J. G., and J. A. Buzacott. 1981. Open queueing network models of dynamic shops. International Journal of Production Research 19(3):255.

Sims, J. R. 1986. Food distribution: A material handling opportunity. Material Handling Users Conference. Atlanta, Ga.: Georgia Institute of Technology.

Skinner, W. 1985. Manufacturing: The Formidable Competitive Weapon. New York: Wiley.

Solberg, J. J. 1977. A mathematical model of computerized manufacturing systems. P. 1265 in Proceedings of the 4th International Conference on Production Research, Tokyo.

Suri, R. 1983. Robustness of queueing network formulas. Journal of the Association of Computing Machines 30(3):564.

Suri, R., and R. R. Hildebrant. 1984. Modeling flexible manufacturing systems using mean-value analysis. Journal of Manufacturing Systems 3(1):27.

Tompkins, J. A., and J. A. White. 1984. Facilities Planning. New York: Wiley.

White, J. A. 1982. The automated factory and integrated systems in the 80's. Proceedings of the 4th IIE Managers Seminar. Norcross, Ga.: Institute of Industrial Engineers.

White, J. A. 1984. Design for automation. Modern Material Handling 39(1):29.

White, J. A. 1985. Selling integrated systems. Modern Material Handling 40(3):29.

White, J. A. 1986a. Becoming the systems integrator by the year 2020. P. 372 in Proceedings of Fall Industrial Engineering Conference. Norcross, Ga.: Institute of Industrial Engineers.

White, J. A. 1986b. Impediments to system integration. Modern Material Handling 41(10):23.

White, J. A. 1986c. Time to re-evaluate your insurance. Modern Material Handling 41(7):25.

White, J. A., and J. M. Apple, Jr. 1985. Material handling requirements are altered dramatically by CIM information links. Industrial Engineering 17(2):36.

Whitt, W. 1983a. Performance of the queueing network analyzer. Bell Systems Technology Journal 62(9):2817.

Whitt, W. 1983b. The queueing network analyzer. Bell Systems Technology Journal 62(9):2779.

DESIGNING AN INFORMATION SYSTEM FOR INTEGRATED MANUFACTURING SYSTEMS

ULRICH FLATAU

ABSTRACT   The development of improved software and the availability of low-cost computer hardware allow hard and soft automation to be combined in ways that can provide powerful complex manufacturing systems. In addressing the system issues, it is necessary to address the problems associated with total integration. This paper describes a process by which complex integrated manufacturing systems can be systematically planned. It discusses how the information system that will support the integrated system can be planned, including anticipating the need to make modifications to accommodate future possible changes in both technology and function.

INTRODUCTION

The recognized decline in the productivity of many U.S. companies over the past 15 years has been a strong stimulant for individual companies to search for ways to improve their operational efficiency and to become more competitive in the marketplace. Many companies have achieved productivity improvements that place them among the most efficient in the world, whereas others continue to search for the reasons for their problems. Although blame is often placed on such issues as government intervention, increases in energy cost, and labor regulations, an analysis of successful companies suggests that a significant part of the problem often results from factors internal to the company. The general approach to management, the concepts used to market the product, the policies followed in making investments, the importance as-signed to product quality, and the reward system for employees have all been shown to critically influence the competitiveness of a company.

The successful companies have changed their attitudes toward customers, their production processes, and their internal management approaches. They have chosen to emphasize high-quality production rather than high-volume production. They have found that the production of higher quality products not only does not cost more money, it reduces costs. By reducing manufacturing costs, it helps achieve higher profit margins and offers the opportunity for the manufacturer to achieve a larger share of the market. In many cases, it has been shown that higher quality products yield as much as 40 percent higher return on investment than do lower quality products. The achievement of this competitive advantage is the challenge that confronts today's managers.

A review of successful companies also suggests that there is no single technical solution that can be easily copied or bought (Hall, 1987; Hayes and Wheelwright, 1984). Quality circles, just-in-time manufacturing, or automation must each be tailored to the needs of the individual company. Managers of manufacturing enterprises face a demanding complexity of individual functions as they attempt to achieve their business objectives. A process, not just a program, is required to achieve corporate goals and objectives. At Digital Equipment Corporation, this process is described as the "system improvement process" or, simply, striving for manufacturing excellence.

Computer-integrated manufacturing (CIM) is the term often used to describe this improvement process. In this paper, CIM is used to mean the entire manufacturing enterprise—not just the production subsystem of a manufacturing operation. Current manufacturing enterprises are complex systems with a multitude of internal and external relationships. These relationships must be optimized if the organization is to achieve the best possible performance.

A principal corporate goal must be the development of clearly defined corporate objectives, including well-defined strategies for product, sales, marketing, finance, and manufacturing. Corporate strategies and functions must be integrated in support of the corporate objectives. The integration of the different strategies and functions in a manufacturing enterprise is critical to its success. Rather than each function attempting to optimize individually its operation, optimization of the overall enterprise must be the primary objective. In addition, the criteria for measuring the performance must be clearly established by management, and these criteria must be consistent with corporate goals and objectives.

SYSTEM INTEGRATION

In considering the manufacturing enterprise, one is immediately confronted with the inconsistencies between the data that describe the product, the technology, or the know-how involved in the manufacturing processes, and the data used to describe the management or the business operation. At least three types of data sets are frequently used in an industrial enterprise. One data set is used in the CAD environment, another for process planning, and a third for manufacturing. All three of these are supposed to describe the same entity and the same object independently. Unfortunately, these data sets are frequently overlapping and incomplete. Only a few companies have been able to achieve a level of integration that allows the use of a single data set.

Integration is a difficult task because of its many different facets (Chestnut, 1967; Synnott, 1987). As everyone knows, integration is not a binary state. A system is neither completely integrated nor completely unintegrated. In fact, every manufacturing system in place today is partially integrated, for without some integration it would not be possible to produce complex products such as aircraft or automobiles. The problems that most frequently arise are with the accuracy of the data, the speed of transmission of the data among different functions, activities, and subsystems, and the difficulty in properly describing entities in the data bases.

In attempting to integrate functions and subsystems into a total system, it is essential first to define the purpose, the boundaries, and the goals for such an integrated system. It is useful to start with the broad system model shown in Figure 1. This input-output model represents a manufacturing enterprise on the highest conceptual level. Inputs of material, energy, capital, and labor are transformed and transported and result in products, waste, and ultimately corporate profits. The system is kept in balance by various feedback and control actions. From an information system point of view, this simple model can be used to represent a manufacturing system that consists

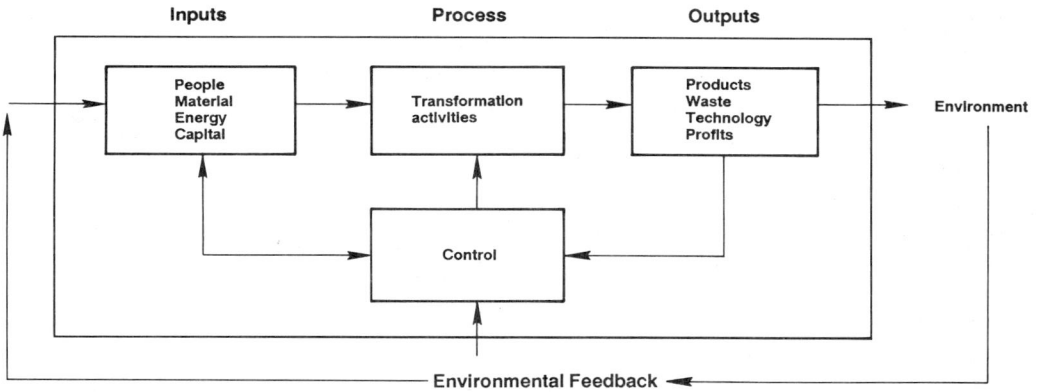

FIGURE 1   Conceptual model of a manufacturing enterprise.

of processes and data, both of which can be handled by a computer. The first step in accomplishing this is to represent the physical processes by algorithms so that appropriate computer programs can be created to control these processes. The second step is to define the data needed by the control algorithms, namely, the data required to trigger processes, to track materials, and to control such items as tools. This model also illustrates how the exchange of data and messages between processes can lead to integration.

Let us take a closer look at what integration means in the context of an information system. Since programs and data make up the system, the degree of integration is determined by an analysis of processes and data. Figure 2 is a representation of the manufacturing enterprise, the broad function needs, and the data and information technology needs. Data in this context must have a name and a value. The problem in many of today's manufacturing systems is that different names and different values have been allowed to exist for the same en-

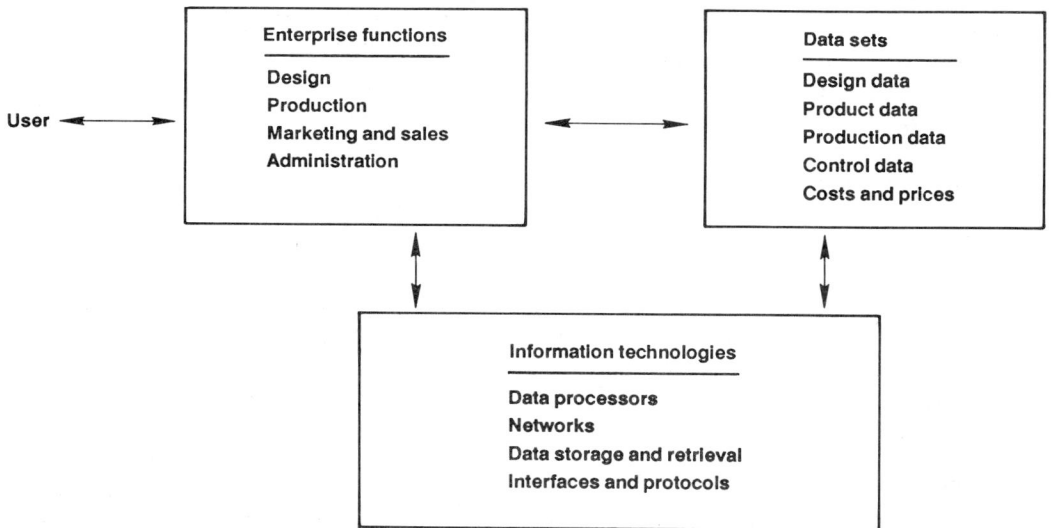

FIGURE 2   Manufacturing enterprise view.

tity. A principal reason for this is that applications for different functions are frequently written by different groups. In addition, most of the current systems are batch-oriented so that timely updates of the data are nearly impossible.

Let us assume that the data name and the data value have been defined for one object or entity. In a distributed system, multiply defined data names and multiple values are allowed for different states of the total system. In a distributed system having two distinct states, a name and value for an entity will possess four distinct values. A logical goal for data integration should be to achieve a single data name and a single data value definition for each of the four entities in the system. Adding the next element, the process, adds two more distinct quantities for each state, thus requiring eight distinct values for the name, value, and process description of the system. Since each of these data values can be stored locally—that is, on the same node—or remotely—that is, on different nodes—another factor of two in the values is introduced, bringing to 16 the different values required if integration is to be achieved. When control is added to the matrix, we are confronted with a system that has 32 different values and thus 32 different stages that must be included in a total integration of the system. The degree of integration must be defined objectively to reduce ambiguity to the maximum degree possible.

Meaningful comparison of differences in the stages of integration requires establishing a set of criteria that reflect the system goals. System integration can be measured by the following criteria: performance, availability, maintainability, extendability, flexibility, system development time, cost of ownership, data integrity, and reusable code. The stages of integration and these criteria provide an objective way to measure and to judge system development alternatives. This can be illustrated by considering the implications of specifying a centralized system, namely, a system with a single data definition, single value definition, and a singly defined process. Although such a system may operate satisfactorily, the performance is likely to be low and the costs high. If the resulting performance and cost cannot be tolerated, it will be necessary to select other criteria that will best support the business goals.

Defining the integration criteria enables the designer to measure the level of integration and to determine the value matrix in accordance with the attributes that will be obtained. These matrices help define the stages or levels of integration by assigning the costs and benefits and by helping develop a plan for implementation.

Integration Criteria

The starting point for the design of an integrated system is the ranking of the integration criteria and the selection of the proper level of integration. If a very short development and implementation time is demanded, one will probably be forced to settle for a multiply defined data and process specification. The reason for this is that the application packages will likely come from different vendors, who will not have the time or incentive to coordinate to the degree needed to accomplish a high level of integration. Nevertheless, a reasonably integrated system can be achieved if the right data and process update strategy are selected at the outset.

Achieving a certain level of initial integration is possible, but keeping an evolving system at that same level of integration is difficult. Ongoing change is difficult to handle if the level of integration is to be maintained. The impact of this process on ongoing system costs is important, since the cost of maintenance is normally much higher than the cost of installation.

System integration improves the capability for achieving good communications within the system. There are three funda-

CIM – System

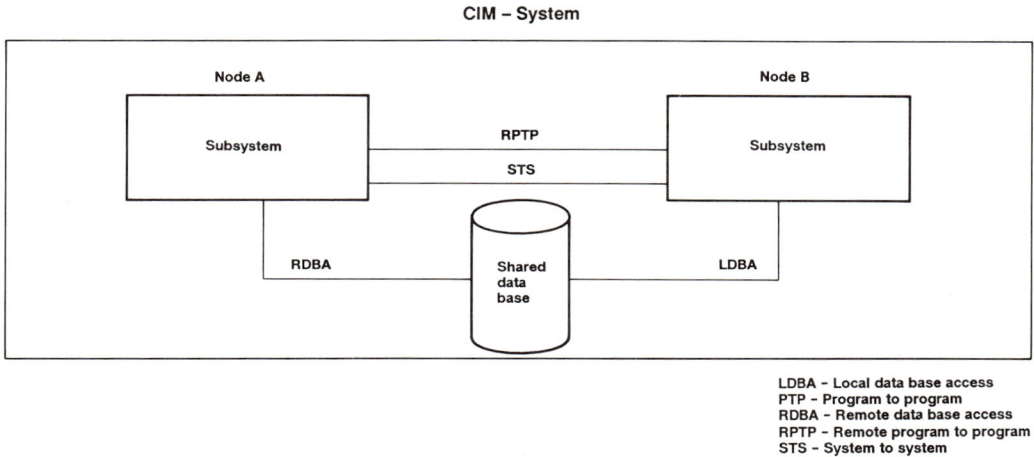

LDBA – Local data base access
PTP – Program to program
RDBA – Remote data base access
RPTP – Remote program to program
STS – System to system

FIGURE 3   Physical data exchange methods.

mental ways of communicating in distributed systems. As shown in Figure 3, there is system-to-system, program-to-program, and program-to-data communication. These fundamental means of communicating can be implemented in three different modes: the batch or buffered mode, the segmented or partly buffered mode, and the real-time mode with immediate response.

Implementation of these different methods of communication requires specialized tools. Hardware, system software, and special application languages are needed. If these basic forms of communication are not supported in languages, operating systems, and hardware, it will be especially difficult to implement an integrated system. The alternative is to write libraries in different languages to compensate for the functional deficiencies and language extensions.

The different forms of communication can be approximately divided into three broad classes. The first is process-to-process communication, which, as the name implies, is a one-to-one relationship. The second is a many-to-one relationship in which many processes talk to one process. And the third involves one process communicating with many others, a one-to-many relation-

ship. Each of these relationships presents a unique set of issues.

In most distributed integrated systems, a single process must be in communication with, or be able to receive messages from, many other processes. Thus, the one-to-one relationship is not generally appropriate for the distributed environment.

Implementation of the many-to-one relationship is severely limited by currently available computer hardware. Most current systems are Von Neumann computers, and they can execute only one operation at a time. Therefore, with most available computers, there is no way to achieve real-time parallel communications.

The one-to-many relationship is limited by system timing considerations. There is currently no way to update the data at different locations in a single time frame. One needs many time frames to distribute data in a one-to-many relationship. Therefore, it is difficult to update data in distributed data bases in a circumstance in which multiple data definition and multiple process definition has been allowed. This difficulty leads, therefore, to a concern about maintaining data consistency when it cannot be ensured that the time delays in updating have not created a set of inconsistent data through-out the system. Current hardware limita-tions need to be critically examined.

The exchange of physical data can help achieve a form of integration. Having a process running on Node A and another process running on Node B requires system-to-system communication. One needs sup-port in the form of language, operating sys-tem, and the network to implement this form of communication. Support for re-mote program communication—that is, di-rect communication—may be needed. A buffered communication for remote data base access or local data base access may be used. All of these different forms of com-munication need to be supported by lan-guages, by operating systems, and by net-works. It is not uncommon to find that the support for the network and the capabilities of the operating system are insufficient to ensure that CIM implementation will be successful.

Subsystem Needs

The next step is to look at the communi-cation needs of the subsystems. Figure 4 shows that in the subsystems program-to-program communication is needed on the same node as well as on different nodes. Communication in this sense means ex-changing data, exchanging commands, and exchanging events. Since real-time data are not stored, direct memory access may be needed in real time to both the local sys-tem and the remote system. All of these different communication functions must be functionally embedded, implemented, and supported by programming languages, op-erating systems, data management systems, and networks. Although this appears to be straightforward, few operating systems al-low for this range of functionality.

There may be needs for access to local or remote data bases in a real-time environ-ment. The simplest task, of course, is ar-ranging for a single process to use its own data. One might also have local data base access, in which case the process and the data are located on the same node or two different applications share the same data base. One may also have remote data base access where the data may be somewhere in a remote system. Obtaining access to a remote data base is very tricky using a net-

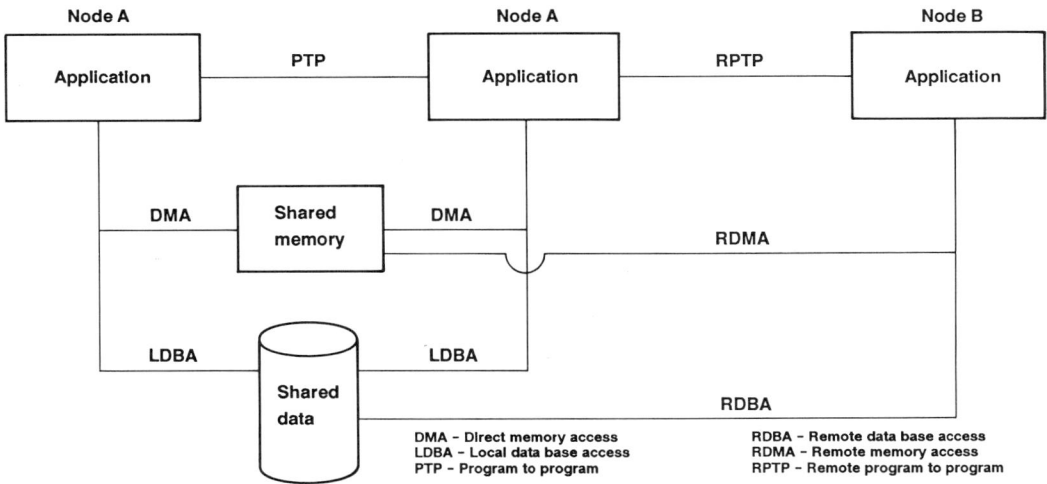

FIGURE 4 Communications needs of subsystems.

work and a programming language. These are the basic prerequisites for communication or integration of distributed systems.

Data Consistency

Even this simple conceptual view of system integration shows how difficult system integration may be and the nature of the tools that are needed for integration. These aspects become even more complicated when one includes the aspects of real-time events, command translation, remote data base access, and control data. There may be many good reasons to accept an integrated system design with multiply defined processes and multiply defined data names and data values, but it is essential that the integration strategy include a procedure by which data and process consistency is ensured.

There are two basic philosophies to achieve the desired consistency—a pull system, which requires the receiver to ask the sender for data, or a push system, in which the sender initiates transfer to the receiver. In a push system, the new data and processes will be automatically distributed to all receivers, as occurs for an electronic mail system with a distribution list. This approach requires considerable control and system intelligence. In a pull system, the user is responsible for requesting the new data and programs before program execution is initiated. Examples of this approach in manufacturing systems are cell controllers and area controller links.

System Planning

It should be clear that integration requires a great deal of system planning. As an ongoing process, it requires a long-term commitment from high-level management, political unity and coordination of all elements in the enterprise, and, finally, strict discipline in function, data, and process

definition. System integration requires some of the same approaches required in the building of a house. For example, serious construction on a house would not be started without a plan. Although such a sequence of events would not be expected to occur in computer-integrated manufacturing systems, systems continue to be started without a clear plan having been developed for the integration. All too often a detailed plan is replaced with a hope that somehow, sometime, all the disconnected pieces will come together to make a highly effective system. What must be made clear is that there are too many different technologies, functions, and requirements in such a system to be left to chance. It will not work effectively without a plan and the commitment from all who are involved.

The engagement of the necessary partners in achieving CIM solutions requires a clear strategic plan and a strategic partnership among all those concerned. The elements of this strategic partnership include

- Integrated planning and implementation
  - Computer technologies
  - Factory automation
  - Office automation
  - Education and training

The creation of a system of this complexity must be viewed as much more than simply writing an order for procurement. It involves the integration of the entire manufacturing enterprise. Many partners have to be involved if all the different technologies, processes, people, and data are to be successfully brought together to form an integrated system. Not only are there no turnkey CIM systems available today, but also, in view of the complexity of the entire system, it seems unlikely that turnkey CIM solutions will be available in the foreseeable future. This suggests that an architectural system approach is essential.

ARCHITECTURAL FRAMEWORK FOR CIM TECHNOLOGIES

CIM systems must be considered as belonging to a class that can be described as human activity systems. Arriving at a clear definition of the problem in human activity systems is difficult. Identifying and solving a problem in physical systems may take considerable time and effort, but experts can usually reach an agreement on the statement of the problem. This is often not the case in social systems. In most social systems, there is no simple problem statement, and there is no simple right or wrong answer to that problem. There is just a problem area.

The information that must be distributed in a human activity system indicates that two separate information systems must be considered. One is implemented through policies and processes (that is, the formal information system), whereas the other (the informal information system) ties the people in the organization together. Many current manufacturing systems still depend heavily on informal information systems. If something goes wrong with the formal system, we still need experienced people to fix it. One also sees the requirement for having an expert with many years of experience with the process, since a mathematical model of the process is not sufficient to allow analysis of all likely problems. People provide the learning capability and thus enable improvements to occur in the operations.

The currently available system analysis and design tools were developed for and by engineers to analyze and to plan complex physical systems. They do not, however, take the total system into account. In particular, the human elements of the system are not represented in these methodologies. Furthermore, these techniques concentrate more on analysis than on design.

In the effort to create effective analytic tools, there has been a movement toward specialization. Division, classification, separation, partition, and segmentation have characterized analysis. Much of the research in each of the separate fields of systematic knowledge has resulted in a further narrowing of the specialization. To create a successful system that properly relates to the human system will require that more time be spent on synthesizing and assimilating the existing knowledge.

Accomplishing transformations is basic to human activity systems. The input-output model is a powerful tool in representing these systems, in that it can be used to describe many different aspects of the system. Since a transformation process or activity can be executed by people or by machines, it is essential that the division of labor between humans and machines be clearly considered. The technical systems, composed of machines and processes, though manageable and predictable, are relatively inflexible. By contrast, the human part of the system is somewhat unpredictable but flexible. Because CIM systems should be highly integrated and flexible, it will take careful analysis to determine the part that should be included in the technical system and the part that should remain in the social system. To ensure that the system will be acceptable to the user, the planning and implementing processes for complex CIM systems must include those users.

An organized way to plan and implement complex systems must be devised. Digital Equipment Corporation's CIM system architecture will be used as an example of how this problem can be approached (Digital Equipment Corporation, 1986). The balance of this paper concentrates on the formal system that relies on procedures, protocols, and detailed specifications for its success.

Distributed Systems

Manufacturing systems are distributed systems. They use distributed resources such as hardware, data, events, and protocols. Their success is determined by the capability to control the processes that are used in the transformation of the resources. Control of the processes involves monitoring the status of processes as well as the management of the resources.

These systems use resources and processing software from different vendors. In this sense they are heterogeneous systems. To integrate these diverse resources it is necessary to use a variety of protocols, interfaces, handlers, and management tools. Many of these tools, however, are proprietary to the vendor that created them.

This discussion leads to the following rather simple characterization of manufacturing systems. They are distributed, heterogeneous, and vendor-supported. It is the combination of these characteristics that led the Digital Equipment Corporation to advocate "open systems."

An "open system" is defined as one that allows the "open" exchange of information among elements of the systems through the use of common standards. For such systems, the exchange of data, either structured or unstructured, files, events, commands, and application processes can be accomplished without regard for the particular vendor that supplies the subsystems.

Architecture for Distributed Open Heterogeneous Systems

Proceeding with a successful development requires rules, blueprints, and guidance for system planning and process implementation. This will provide a framework for systematically examining application development and selection, the integration of product development, the coordination of vendor developments, and the planning for services such as training. Having a plan in place is especially desirable for product development. If product developers agree on standards, the development of subelements can be done in parallel, thereby shortening

the implementation time and reducing the cost of the system. Systems can be implemented one piece at a time, with confidence that system extension is possible and easy to accomplish. A successful plan for distributed, heterogeneous, and open systems is critically dependent on standards.

Thus, the need for a plan is clearly dependent on the creation of a system architecture. In abstract terms, an architecture is the structure and relationship among system components. It is a framework for logical and functional implementation that uses the same rules and principles for product design as are used for system implementation. The steps that must be taken in creating an architecture are

- Establishment of goals;
- Identification of hypotheses and models;
- Establishment of principles and rules;
- Integration of available resources; and
- Identification of the necessary technologies.

The goals for a system architecture for discrete manufacturing must be to provide growth paths that build on investments in existing CIM systems and to provide a structure for CIM system optimization. The fundamental hypothesis in the creation of the CIM system and the hypothesis around which the architecture is developed is that all processing functions and related control functions are expressible in the form of data and that these data can be transferred, processed, and stored by computer technology. To be useful, a CIM architecture must consider a common language, system definition, reference models, standards, and principles and rules for system and product design. Enabling technologies, techniques, operation policies, business rules, and so forth will change. Therefore, the architecture should allow for an infrastructure that anticipates that changes will be needed.

Planning a CIM system requires that answers be developed for these basic questions: Why does the system exist? What is it going to produce? How will the product be produced? Because we do not often start from scratch, we usually take existing manufacturing systems and define and plan how to improve them. It is particularly helpful in the planning phase to have a conceptual model of the system in question. The components of the system model are people, processes, data, automation technologies, and material; the relationship among these components may be organizational, functional, logical, temporal, or physical. Timing relations deserve a special comment because they are so complicated. Information is valuable only when one receives it at the right time. This is one of the key goals of CIM systems—providing the right information at the time that it is needed.

Although one of the CIM system hypotheses is that the movement and transformation of material can be triggered and controlled by data, this is not explicitly discussed here. White (1988, in this volume) discusses this aspect of CIM systems.

CIM systems are business-oriented human activity systems that recognize that a minimum set of activities or processes must be executed to produce a product. There is also a minimum set of data that ties the different activities together. Therefore, the approach to system analysis and system synthesis is a business-based system approach, where the functional model reflects the business, its processes, and its data. This model will change only when the nature of the business changes. Since this is infrequent, the functional system model should be very stable. The system objectives dictate necessary activities or processes in a manufacturing system. Based on this assumption, the answer to the often-asked question "What comes first, data or processes?" is clear. The sequence that should be followed is

1. Identification of the business function;
2. Definition of the logical data that are needed;

3. Definition of the physical data that are needed; and
4. Definition of the physical functions.

During the process of analysis and design, it is essential that the user and the system people be brought together. It is useful to encourage the data structure definition to be done by the system people in a top-down approach, with the activity and process definitions being done in a bottom-up approach. The user must be the one to identify what is needed to be successful. But the user is normally not an expert in information technology, and thus the data structure should be designed by system people. After both groups have completed their tasks, there should be a cross-check of whether all the data are available for the defined processes and vice versa.

Before applying specific techniques such as structural analyses and design techniques to identify the system activities and the data needs, one must first define the boundaries of integration. Then it is necessary to decompose the system into relevant functional subsystems and to define the basic business functions and entities. From there, it is necessary to define, in parallel, the data structure and the business subfunctions. This procedure allows the development of any level of detail in the analysis.

A functional model, on the highest conceptual level that reflects the business function of the desired system, is useful in helping define the integration boundaries and in answering the question, "Where should we start to integrate?" According to the predefined business goals and objectives, one should select a functional subsystem and start system integration where it might be most desirable and successful. Although it is desirable to design subsystems so that they can work together, it is entirely impractical to plan or understand simultaneously all the subsystems that an enterprise will need to run the business. What is needed instead is freedom for user areas to employ their own

Corporate Goals and Objectives

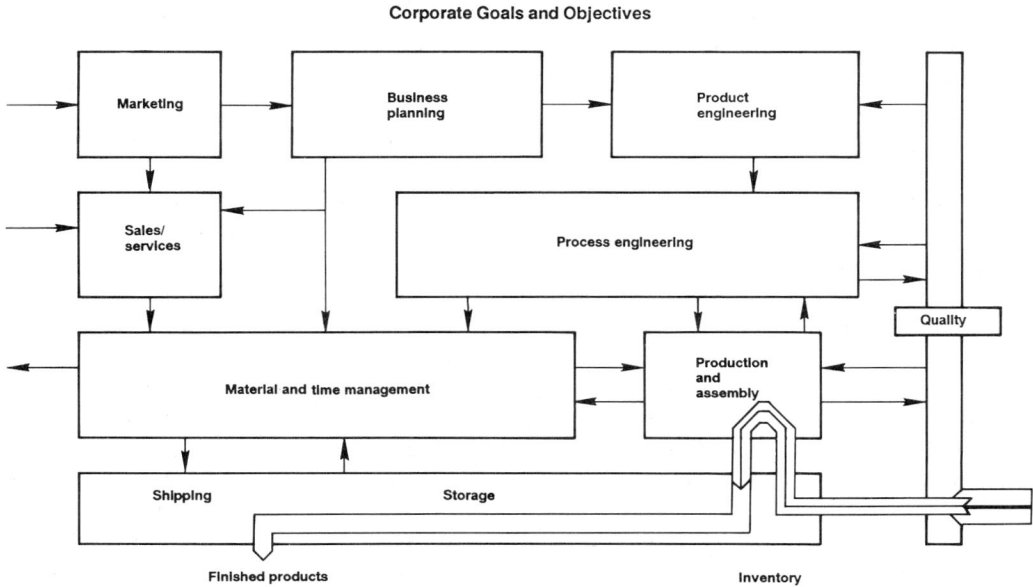

FIGURE 5 Functional diagram of manufacturing enterprise.

subsystems and at the same time to insist that they obey certain rules. These rules need to establish a stable framework for selection or development of an application and to set the conditions for data exchange in such a way that independent subsystem development can proceed.

The outcome of an analysis of the form described can be demonstrated with the help of Figures 5 through 7. There would first be created, as in Figure 5, a high-level conceptual model of the manufacturing system. Second, as shown in Figure 6, a more detailed description of the subsystems would be developed—for example, the product design function. Figure 7 shows the information flow. On the next level of functional analysis, each of the entries in Figure 7 would be expanded for each specific task. An example of such a task might be structural analysis. In this example, the next level of detail would include the logical connection between topics such as thermoanalysis and static analysis. The data flow for a specific subsystem must indicate the data created and when they are updated or used by the function.

With the ongoing activities and the data flows having been identified, the layout of the physical system can be accomplished. An integrated functional model, both a data model and a physical model for the selected business subsystem, should be the result. It is important to note at this point that the integrated system that has been created can be entirely manually based. There is no need to use computer technology to support such a system. The principal reason for using automation and information technology is to increase speed and performance so that the information and goods—e.g., the material and tools—will be delivered on time. Also, an automated system should improve the accuracy or quality of data over those systems in which data are normally reentered.

After the task to be accomplished has been defined, it is necessary to answer the question, "How can it be done?" One goal of the architecture is to make independent, to the extent that it is possible, the business functions from the technology functions. As noted earlier, the business model changes slowly for a company, whereas the technol-

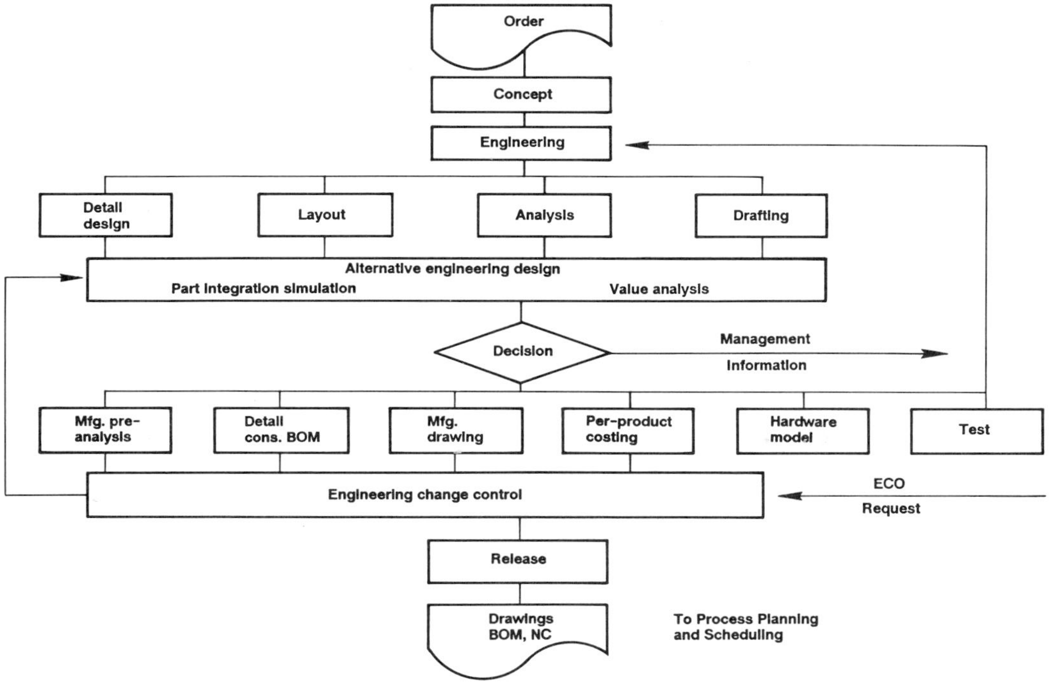

FIGURE 6   Product design functions.

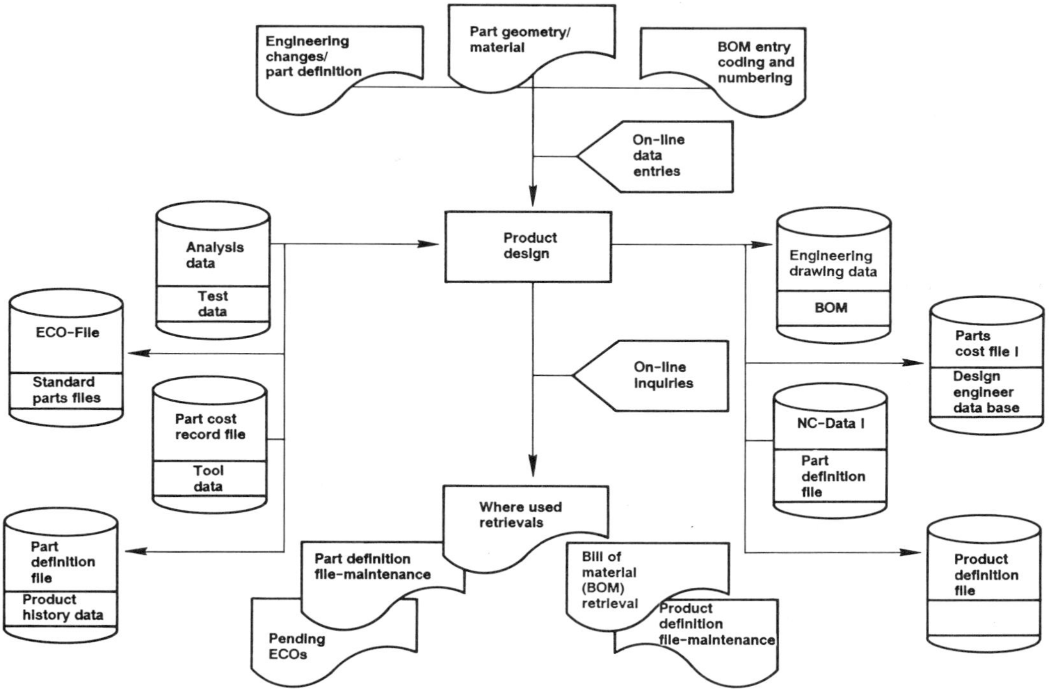

FIGURE 7   Information flow, file level.

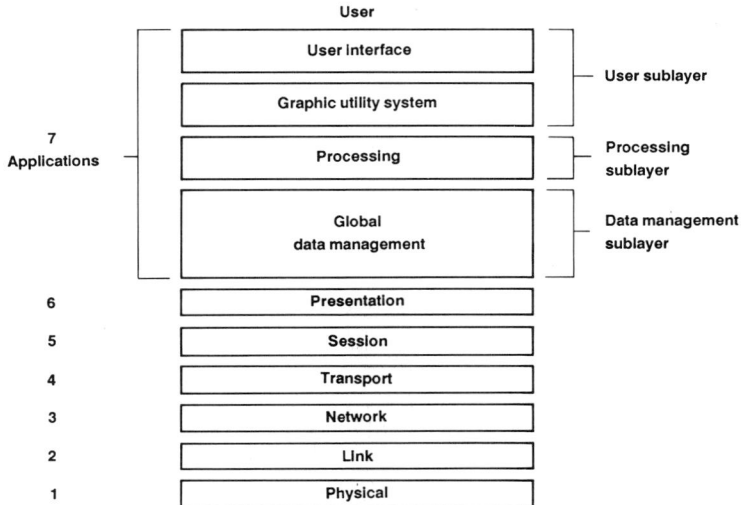

FIGURE 8   Proposed or available standardized interfaces and protocols
for implementation of computer-integrated manufacturing.

ogy used to support the business activities may change at any time. Thus, it is desirable in the architecture to separate function from supporting technology.

Functions can be grouped by different criteria. For example, functions can be grouped according to the same supporting technology or according to their use of the same or similar data. According to the simple process model being used here, only four major layers are considered in the technology model. There is the data creation or use layer (data sink and data source), the processing layer, the data storage and retrieval layer, and (for distributed systems) the networking layer.

The best way to separate distinct functions is by layering, as shown in Figure 8 (International Standards Organization, 1982). Layering allows description of the desired capability, independent of the technology used in the software or hardware. All layers are independent of one another. Although they use the functions, it is not important how that function is performed. One of the best-known layered models is the open-system-interconnect (OSI) model for networks. The proposed model shown

in Figure 8 is an expansion of the OSI model in that it includes the data management sublayer, the application sublayer (in this case, processing), and the user interface sublayer. It also includes integration of resources (e.g., interfaces, handlers, and protocols), and it shows how they are used to separate layer and sublayer from each other. Using this model allows us to describe the desired functionality from one layer without defining how it should be implemented.

Each layer contains functions that are manifestly different in the process performed or the technology involved. The purpose of layering is twofold: first, to take advantage of advances in hardware or software technology without changing the service expected from and provided to the adjacent layer and, second, to hide from the user everything that is not directly related to the user's function. Interfaces are rules for the physical communication between dissimilar layers or sublayers. The interfaces handle communication between dissimilar layers, provide access to service from the lower layers, define rules for service access from higher layers, and handle the process of exchanging data among layers. Handlers define the connection between the service layer and the supporting operating system and hardware. Their purpose is to provide access to operating system service, to provide access to device drives, and to decouple service layers from the hardware. In this sense, a language compiler can be thought of as a handler. The user layer allows device-independent data input and output and user communication with the processing layer. It also provides the user an interface with the system service. The processing layer provides resources for process execution. The processing layer supports the execution of system management processes, application processes, and application management processes. It also allows applications to exchange data automatically and to provide for message exchange between heterogeneous applications. The data management layer provides data and device independence for data storage and retrieval, and it provides all necessary functions for data and file management in open-distributed and heterogeneous systems. It also ensures data

consistency, security, and accuracy in distributed and heterogeneous systems.

For horizontal integration, networking tools are needed. Networking provides physical and logical connections to other open systems. It provides for data, file, and command exchange between distributed systems.

Horizontal integration in distributed systems requires a broader tool-set. Protocols are required. Since the protocol defines the communication rules between similar service layers, it is only a logical concept. There are three fundamental functions the protocol performs: establishing a necessary convention, determining a standard communication path, and identifying a standard data element. Protocols are required for networking in distributed systems and for stable application-layer integration and connection to other open-distributed and heterogeneous systems. Not only do they provide for data file and command exchange between different systems and applications, they are necessary to understand the meaning of data transported between applications.

By using standard protocols, it is possible

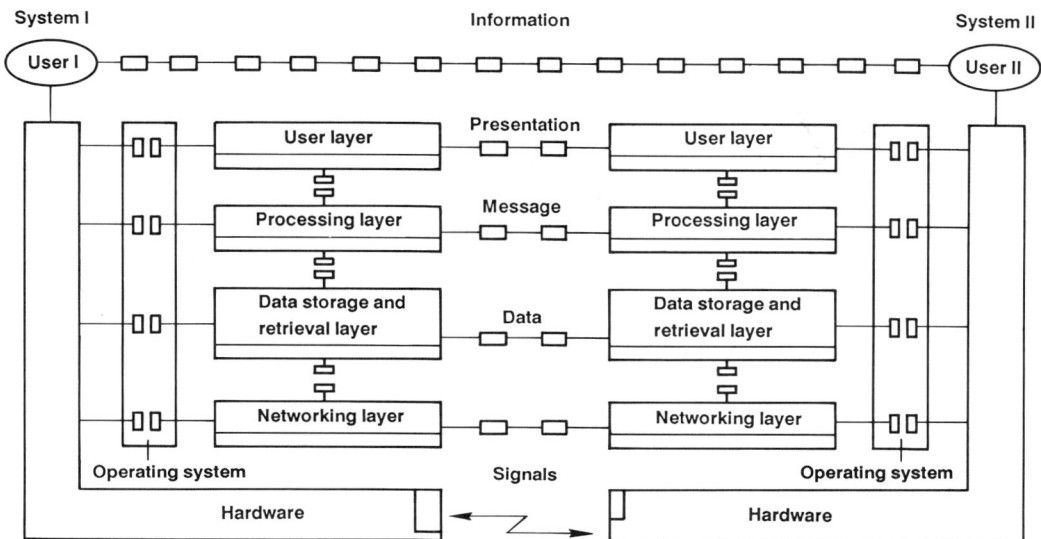

FIGURE 9 Layered model of open system architecture for computer-integrated manufacturing.

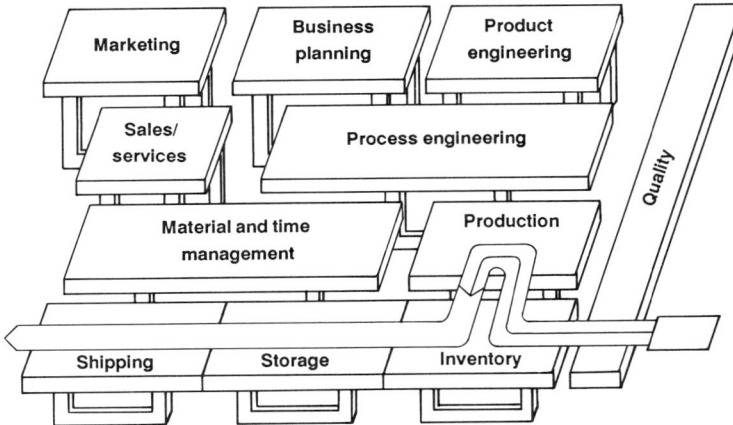

FIGURE 10   Enterprise function supported by information technology.

to reduce dramatically the total number of protocols that are needed for system integration. The model shown in Figure 9 clearly indicates how one can separate function from technology, and where one needs protocols, interfaces, and handlers to integrate heterogeneous systems. It is obvious that interfaces and handlers can be proprietary to achieve integration and data message presentation, but protocols must be standard.

Every activity or process that was identified in the earlier functional analysis is a candidate to be supported according to the technology model. Manually supported activities and automated technology-supported activities can coexist in systems that are planned and implemented according to the architecture described in this paper. One can automate and integrate a subsystem and later integrate this subsystem with other subsystems, as shown in Figure 10.

Generally speaking, there are three ways to integrate systems. The first is to design, plan, and implement without regard to any existing system or physical entity. Seldom does one have the chance, however, to build an integrated system with a "clean piece of paper." The second approach is to extract data from existing systems to build an integrated system. A rare case occurs when data can be used "as is"—that is, when an exist-

ing system satisfies function and data models. The third and most usual way is to design and implement a system in pieces over a period of time. This means replacing the old system with one that is new and more integrated. Such a migration does not contradict the goals of integration when the conceptual model for the total system is well defined. It is not possible, however, to retrofit a high level of integration into a system where the business function, analysis, design, and concept of data flow have not been included from the beginning.

The conceptual model of the manufacturing enterprise presented here allows the grouping of subsystems into logical business functions. These different subsystems—administrative operations, graphical interactions, number crunching, batch processing, transaction processing, and real-time control—represent different user groups with unique requirements of the supporting information technology. The users of these subsystems execute administrative operations, work with interactive graphics, execute CPU-intensive processes, work on transaction processing systems, and operate in the real-time environment. To reflect the different user needs, we have defined different user profiles, using the general model shown in Figure 11. Each of these profiles

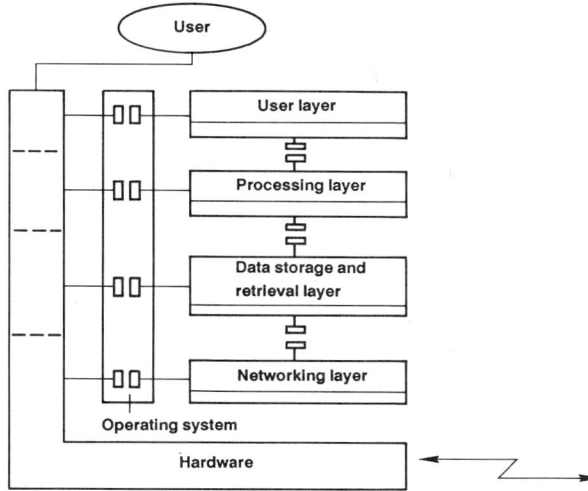

FIGURE 11 System architecture for computer-integrated manufacturing: Layered model.

(Figures 12 through 15) shows a different implementation of the same functional layers.

The unique communication, data management, and operating system requirements for a real-time environment lead to a different implementation in the different functional layers. It is obvious that there is not one single application protocol or networking protocol that fits all user profiles. The results show that the requirements for hardware, operating system, user interface, and data management differ from user profile to user profile. In planning for the

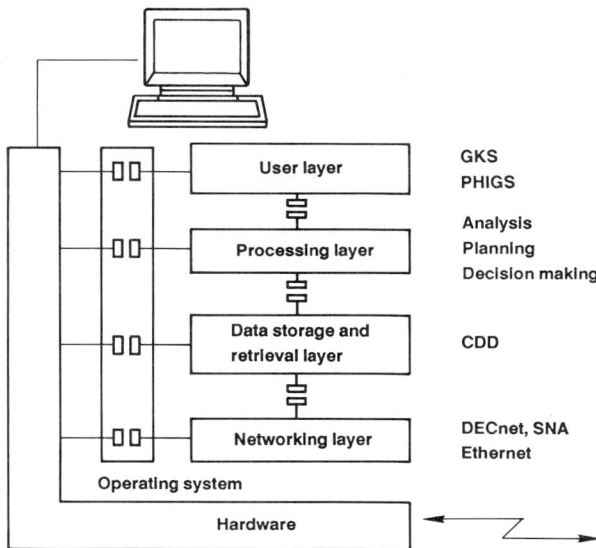

Processing Support • Transaction Processing • Report Generation • Inquiry Processing

FIGURE 12 Planning and control profile.

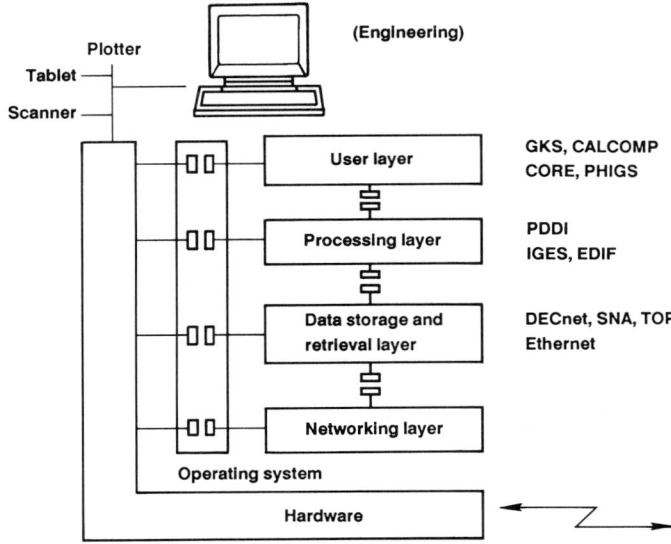

Processing Support ● Number Crunching ● Matrix Operations ● Interactive Input/Output

FIGURE 13   Interactive graphic profile.

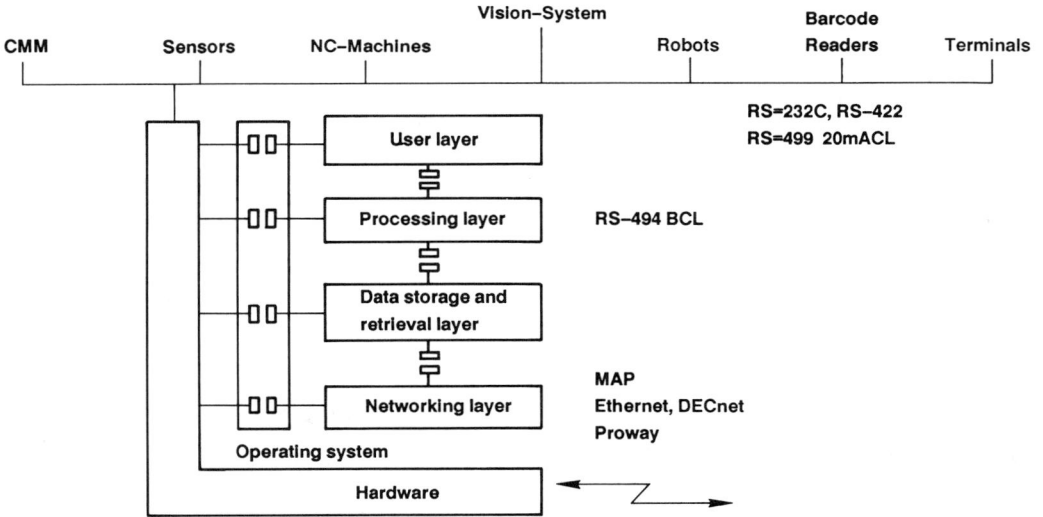

Processing Support ● Logic Operation ● Scanning ● Interrupt Support

FIGURE 14   Real-time user profile.

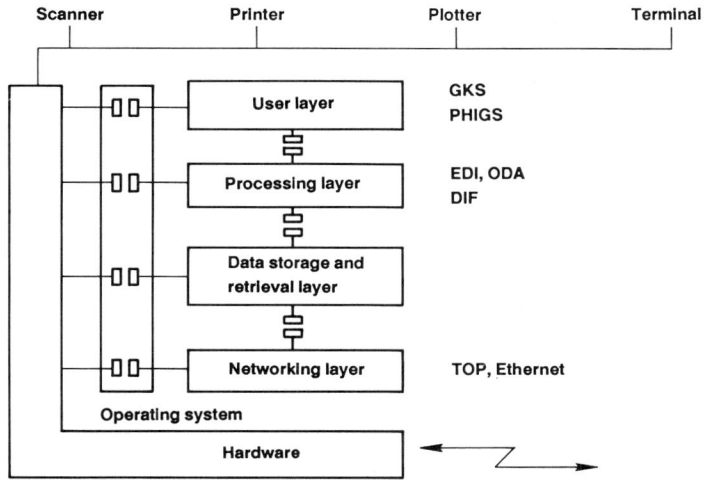

FIGURE 15   Office profile.

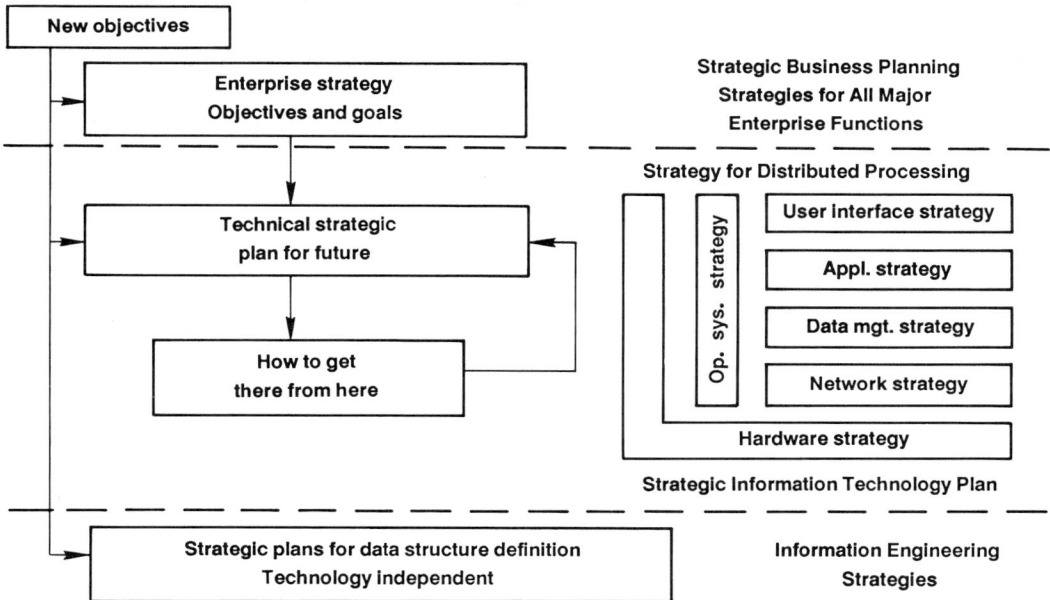

FIGURE 16   Strategic planning for computer-integrated manufacturing.

implementation of an integrated system, it is essential that one select the right infrastructure for all five user profiles.

CONCLUSION

Many people believe that integration can be achieved by simply developing a networking strategy. Others believe that a unified user interface brings integration. This paper suggests that it takes much more to achieve integration. Management must develop a business strategy, and this business strategy must be supported by a technical strategy that includes production process strategies and strategies for distributed processing. Management must agree on strategies for hardware operating systems, user interfaces, application data management, and networking for all five different user profiles. Therefore, management must take the lead in developing a strategic information technology plan. Figure 16 is offered as one example of a strategic plan for CIM that incorporates all of these various attributes.

REFERENCES

Chestnut, H. 1967. Systems Engineering Methods. New York: Wiley.

Digital Equipment Corporation. 1986. CIM-Architecture Version 1.1. Marlborough, Mass.

Hall, R. M. 1987. Attaining Manufacturing Excellence. Homewood, Ill.: Dow Jones-Irwin.

Hayes, R. H., and S. C. Wheelwright. 1984. Restoring Our Competitive Edge: Competing Through Manufacturing. New York: Wiley.

International Standards Organization. 1982. ISO/TC 97: Information Processing Systems. Draft International Standard ISO/DIS 7498.

Synnott, W. R. 1987. The Information Weapon: Winning Customers and Markets with Technology. New York: Wiley.

White, J. A. 1988. Material handling in integrated manufacturing systems. In Design and Analysis of Integrated Manufacturing Systems, W. Dale Compton, ed. Washington, D.C.: National Academy Press.

INTEGRATION AND FLEXIBILITY OF SOFTWARE FOR INTEGRATED MANUFACTURING SYSTEMS

ARCH W. NAYLOR AND RICHARD A. VOLZ

ABSTRACT   This paper suggests a new, integrated approach to research on real-time control software for integrated manufacturing systems. The approach is based on five assumptions about how control software should be designed and developed. They are (1) that control software should be written as assemblages of software components, (2) that this writing should be done in a largely common distributed language and associated software environment, (3) that explicit formal semantic models are required, (4) that generics will significantly expand software reusability, and, finally, (5) that these concepts are entangled and should not be researched separately.

INTRODUCTION

In this paper we discuss two things, both of which are prescriptions for manufacturing research. The first might be described as a general goal for manufacturing research. The second is a research approach that moves, to a limited extent, toward this goal.

One must be somewhat humble when attempting to prescribe a plan for U.S. manufacturing, even if only for the research. Over the past 5 or 10 years so many medicine men have touted their magic cures that audiences and readers have become a bit skeptical. We have had just-in-time delivery, quality circles, robots, computer-integrated manufacturing, Japanese industrial democracy à la Fremont, yen-dollar realignment, and so on. Perhaps the most popular cure—one that any eighteenth century physician would recognize—has been

good old-fashioned bloodletting. But these cures have not been found completely successful, thus raising the possibility that new approaches are needed.

The difficulty is, of course, that manufacturing is an extremely complex activity with intertwined social, economic, organizational, and technical aspects (Volz and Naylor, 1986). Unfortunately, studying "complex activities with intertwined social, economic, organizational, and technical aspects" is difficult. Even with the best of initial intentions, we almost invariably end up following the path of reductionism and studying one aspect of the problem from the viewpoint of one intellectual discipline. The usually unspoken assumption is that somewhere else, at some time in the future, somebody else will bring it all together.

Indeed, we follow this path even with purely technical problems that are complex. Research is usually concerned with

79

specialized techniques, not complex applications. Perhaps this is unavoidable, but one
does have the nagging worry that it will
never "all be brought together" and that
practitioners will continue to grapple with
the problems as best they can, without the
aid of meaningful foundations. Although it
will not be easy, we believe that "bringing
it all together" should be a general goal for
manufacturing research.

Although we would have liked to introduce a dramatic new paradigm for the integrated study of manufacturing systems,
we do not have one. The proposed prescriptions will be far more modest, but at least
they will be an attempt to move in the right
direction. The discussion will deal with
an integrated, mildly nonreductionist approach to one technological aspect of integrated manufacturing; in particular, it is
concerned with the design and development of real-time control software for the
control of the flow of parts, tools, material,
and information in a computer-integrated
manufacturing system. The approach is
nonreductionist in the sense that major aspects of the problem are confronted simultaneously in a unified manner.

This paper first describes the systems of
interest in some detail. This is followed by
a review of how control software is currently being realized, and the limitations of
current practice are highlighted. The bulk
of the paper presents the proposed approach to research on control software and
an example of an application of this ap-

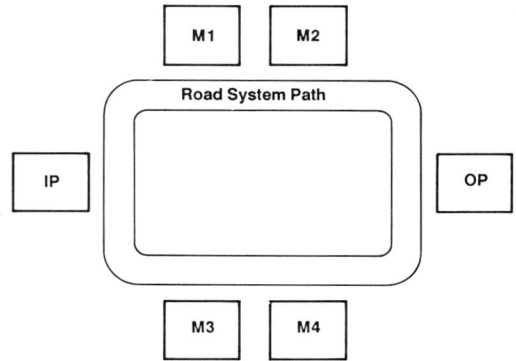

FIGURE 1 Mechanical perspective of the manufacturing system.

proach. Finally, some proposals are made
for the research questions that result from
this approach.

SYSTEMS OF INTEREST

A rather simple, and undoubtedly familiar, version of a system of interest is shown
in Figures 1 and 2. Figure 1 shows the physical layout of various devices, that is, numerically controlled (NC) machines, measuring machines, robots, and so forth. The
backbone of the system from this perspective is the material transport system. It provides the physical integration of the system.
Figure 2 shows the same system from the
perspective of information and software.
Here the integrating backbone is the local-
area network. The device controllers and
the control computer are connected to it.
More elaborate systems might have more

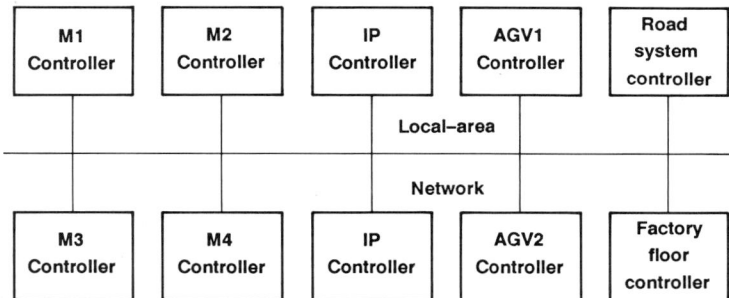

FIGURE 2 Software perspective of the system shown in Figure 1.

devices, more device controllers, more material transport systems, more local-area networks, or more control computers. This simple version will suffice, however, for the present discussion.

A PERCEPTION OF CURRENT PRACTICE

A few years ago a discussion about current practice would probably have focused on "islands of automation" and agonized over the difficulty of making low-level interconnections. It might have been noted that the interfaces for NC machine controllers were either poor or nonexistent, and surely there would have been comments on the lack of standardization in local-area networks. Today, such problems, although not behind us, are recognized, and progress is being made. For example, manufacturing automation protocol (MAP), a local-area network standard for manufacturing (Kaminski, 1986), is being widely accepted. Thus, it is probably fair to say that, over the next few years, low-level interconnection will cease to be a problem. However, once the low-level interconnection problems are behind us, we immediately come face-to-face with far more daunting problems.

The root problem is that so far it has been extremely difficult to reuse or port manufacturing software. The typical situation is that software is tailored to a particular manufacturing system that is designed to make a particular mix of parts. This is not surprising, for a number of reasons. First, there is a very large number of candidate devices from which a manufacturing system may be constructed, and some of them come with their own languages. Second, the structures of systems—that is, the physical and electronic arrangement of the system—can vary markedly. Third, the execution environment for the software is inherently distributed, and explicit handling of communication presents a surprisingly

heavy programming overhead. Moreover, debugging and software reliability, always difficult because of the size of the programs, become all the more so because of distribution. Fourth, the systems to be controlled are usually stochastic, because of such things as device breakdown and random material arrivals, making the formulation of the needed control algorithms extremely difficult. Since a practical general approach to such algorithms is missing, the algorithms are invariably quite specific.

Consequently, the software is expensive because its cost is difficult to spread over several systems. It is inflexible in the sense that it is difficult to change to make new parts. And its reliability is achieved at a further cost in flexibility—that is, the systems are potentially so complex that flexibility is sacrificed just to achieve reasonable confidence that the system will work.

This, then, is our perception of current practice. It should not be taken as a criticism of those writing real-time control software; rather, it should be taken as a statement of the "software facts of life."

THE APPROACH TO SOFTWARE COMPONENTS AND THEIR ASSEMBLAGES

The balance of this paper is concerned with research on real-time control software. As will immediately become evident, our approach starts with some basic assumptions about how control software should be designed and developed, and therefore the research issues become, on the one hand, an exploration of the consequences of these basic assumptions and, on the other hand, the testing of their appropriateness. The basic assumptions are (1) that control software should be written as assemblages of software components, (2) that this writing should be done in a largely common distributed language and associated software environment, (3) that explicit formal semantic models are required, (4) that generics will significantly expand software reusability, and, finally, (5) that these concepts are entangled with one another and should not be researched separately.

The first part of this approach may not seem terribly revolutionary. Since the problems to be addressed are large, complex software systems, it seems reasonable to prescribe modern software engineering concepts. But beyond this general prescription, it appears that the wide application of one particular concept could have a major impact. The concept comes under labels such as software components, objects (Rentsch, 1982), and abstract data types (Shaw, 1980). Other applicable terms are *data abstraction* and *data hiding*. We will use the term *software component* (Volz et al., 1986) to describe all of these elements.

It is easy for anyone with a background in systems to appreciate what a software component is: It is the software equivalent of a black box. It has inputs and outputs to which the user has access. Software components are generalizations of things such as subroutines, procedures, and functions. Typically, a software component is a collection of functions and procedures that operate on some common something inside the component, or, if you will, inside the black box. Consider a simple example.

A software component that establishes the public interface for a stack or last-in, first-out queue is as follows:

```
package STACK is
    procedure PUSH(X:INTEGER);
    function POP return INTEGER;
end STACK
```

Although this example happens to be written in Ada, it could just as easily have been written in Modula 2 or C++. The significant aspect of this example is what it makes available to the user, in particular, the procedure PUSH and the function POP. PUSH adds an integer—this STACK stacks up integers—to the queue, and the function POP takes one off. This public interface and an understanding of how last-in, first-out queues work, that is, a semantic model for stacks, is all the user needs to know to use this software component. In particular, the user does not have to know the details of the internal part of the component, and one can appreciate that the stack could be implemented in various ways. Let us imagine a bizarre example. The stack is not implemented in main memory or even secondary storage. Instead, we have a robot that writes integers on a large blackboard in a long line. We also have a vision system that reads these integers, and the robot has an eraser. This admittedly silly example makes two points. First, the user really has access to the public interface only; second, one can imagine manufacturing devices as being inside a software component. The second point is important. We believe that robots, NC machines, material transport vehicles, and so forth should be incorporated into a manufacturing software system in exactly the same way that our simple stack is—that is, through a public interface.

The ability to create and use software components is not simply a matter of programming style. It requires that the language used have specific capabilities. FORTRAN and BASIC, for example, do not have these capabilities. The component must be capable of being compiled separately from the software that uses it, but, more importantly, the interface and internals must be able to be compiled separately. The public interface is like a control panel for the component. It contains the names of procedures and functions together with their associated parameters. The fact that this interface can be compiled separately from the internals of the component means that users can have compilers *automatically* test the compatibility of the component with the system in which they will use it.

Separate compilation opens the possibility of a true software components industry, and that in turn promises a reduction in software costs. If it becomes possible to purchase software components "off the shelf" and incorporate them into larger software systems, this will create competition for producing software components. Indeed, larger software systems will largely be assemblages of components, and these assemblages will themselves usually be software components. Separate compilation is important because the vendor can deliver the compiled component together with the source code for the public interface. That is, the vendor can sell the component without having to reveal the source of its internal implementation. The user, knowing the public interface, will know enough to use the component; in particular, separate compilation of the interface and internals will allow the compiler to check that the component has been integrated into the user's software system in the proper manner. Alternatively, the *purchaser* might specify the public interface, verify that it is compatible with the system in which it will be used, and include that specification as part of the requirements for the component. The interface can become part of the contract between the purchaser and the supplier. This would mean that the internals might be purchased from another vendor, thereby increasing competition.

A software components industry (Volz et al., 1986) becomes particularly interesting when one considers manufacturing devices and systems. In particular, one can, as suggested earlier, imagine treating robots, NC machines, vehicles, and so forth as software components (Volz and Mudge, 1984). Ignoring for the moment the inevitable distributed nature of a manufacturing system, the integration of a robot into the software system becomes merely a matter of calling the procedures and functions in the robot's public interface. In fact, the robot might be part of a manufacturing cell, and this cell might incorporate a number of other manufacturing devices that are treated as software components. Figure 3 shows a simple example of seven software components: cell, machines, transport, two NC machines, robot, and vehicle. The dotted horizontal line divides the purely "software world" from the "mechanical world"—that is, the actual factory floor. The private internals of the component machines contain the two NC machine components plus software for creating the assemblage of these two components. This latter software is

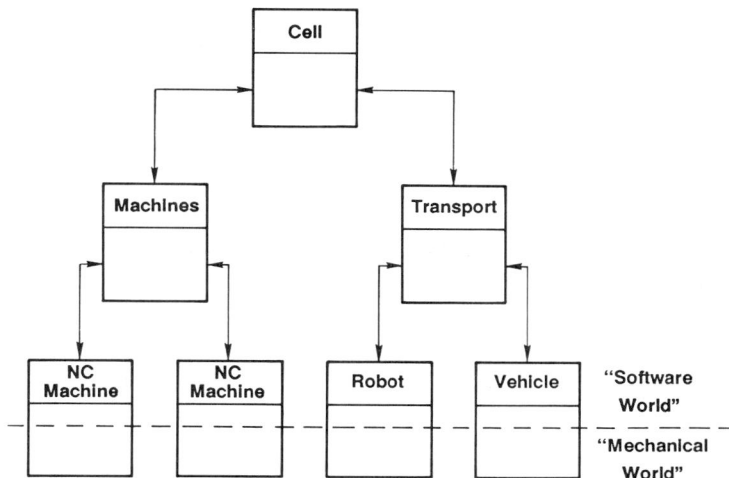

FIGURE 3 Software components of a representative manufacturing cell.

represented in Figure 3 by the bottom half of machines, whereas the top half represents the public interface of machines. Since the NC machines' components span the software and mechanical worlds, the component machines must also. Indeed, each component in Figure 3 spans both worlds for the same reason. The user of cell has access to the other components only to the extent that their function somehow evidenced in the public interface of cell. Similarly, if a number of cells are assembled into a component factory floor, the user of factory floor will have to interact with only its public interface and will not have direct access to any of the cells.

Part of this approach, then, is to use software components as building blocks to create other software components. In a certain restricted sense, this suggests a common language (Volz et al., 1986)—one that supports software components—in which the system integration is carried out. It would be unrealistic, however, to rule out other languages completely. For example, the language used for NC machines is unlikely to change. But this does not preclude a central role for a common language. In particular, if we distinguish between what we call the Euclidean and logical views of manufacturing systems, we believe that using a largely common language is reasonable and desirable.

EUCLIDEAN AND LOGICAL VIEWS

The Euclidean view of a manufacturing system is characterized by locations, orientations, and movements of objects expressed in finite-dimensional Euclidean spaces. Models are typically coordinate transformations, kinematic constraint equations, differential equations, and so forth. Tools come from a variety of sources, including control theory, elasticity, and optics. Typical problems include moving from one geometric point to another, following a path, determining a path, locating a point with a vision system,

or inserting a peg in a hole. Usually the Euclidean view effectively divides the factory floor into largely independent or loosely coupled subsystems—for example, a robot's motion or an NC machine's actions.

The logical view is, as the name suggests, largely concerned with logical conditions and transformations of logical conditions. Variables simply identify things and, except for time, are usually not real-valued. A typical condition might be stated as "the material transport vehicle is at machine number seven." Here the phrases "material transport vehicle" and "machine number" identify two entities in the environment of interest. The phrase "is at" establishes a logical relation between these two entities. Examples of conditions that involve logical quantifiers are "there is something on the material transport vehicle" and "there is nothing on the material transport vehicle." Transformations of conditions involve actions such as moving a vehicle from one place to another or loading a machine with a part. The essence of the logical view is that two things cannot be in the same place at the same time and that it takes time to carry out actions. In addition, the logical view recognizes that events may not turn out the way we thought they would—that is, it allows randomness.

The importance of this distinction between the Euclidean and logical views is that it largely separates local problems from global integration problems. Local problems are typically Euclidean with some logical overtones, but global problems are usually logical. In particular, system-wide real-time control problems must typically be considered logical as, for example, is real-time scheduling. Thus, there are really two kinds of real-time control software: one for controlling the Euclidean view and one for controlling the logical view. Although much of what is proposed here applies to both, the balance of this paper focuses on the logical view.

Returning to the question of the language environment, one can foresee a common language for the logical view and a general way of handling a local subsystem that is programmed in another language. This can be accomplished by encapsulating the subsystem in a software component that externally appears to be written in the common language (Volz et al., 1986). One can imagine encapsulating a robot in this manner.

Distributed Language Environment

There is, however, another important language issue. As was already noted, the software execution environment for a manufacturing system is inevitably distributed. There are three obvious ways to reflect this fact in the language environment.

One way is to program each processor separately, with part of the programming task being the development of required interprocessor communication software on a case-by-case basis. This is typical of current practice and has the obvious disadvantage that writing ad hoc communication software is expensive. Another disadvantage is that debugging and error checking are essentially limited to single processors. That is, the software must be treated as a collection of pieces rather than a single system. Still another disadvantage is that the software is partitioned along processor boundaries, and these may not be the natural or logical boundaries.

An alternative to this approach is based on a distributive operating system. Presumably this would significantly diminish the need to program communications explicitly; however, debugging and error checking would still be fragmented, and so would unnatural program partitioning.

Unfortunately, neither of these approaches adequately supports the concept of software components. The problem is that a software component can extend over more than one processor. For example, the robot component shown in Figure 4 has its

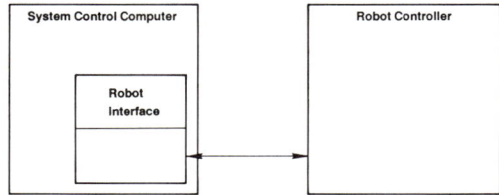

FIGURE 4   A software component spanning two processors.

public interface visible on the system control computer, and most of its private internals are contained in the robot's controller. We believe that the programmer should not have to worry about this separation. In other words, in addition to the language environment being common, it should also be a distributed language (Volz et al., 1987). In this way, it should be possible to write the real-time control software as one program, where a component might encompass several processors.

The power of a common distributed language environment is that it simplifies integration. Since the communication paths are transparent to the programmer, the programmer is allowed to think about the entire program without having to be concerned with processor boundaries. Further, debugging and compile time error checking take place over the entire program. This approach requires, however, a language translation system that is more than just a conventional compiler. Such a system is currently not available. Indeed, distributed languages are currently an active area of research.

Our research on language translation systems has focused on a distributed version of Ada (Volz and Mudge, 1987; Volz et al., 1985, 1987, 1988). Ada was selected for various reasons, including the fact that Ada will be widely used and supported. This is important because manufacturing software must be in the mainstream if it is to benefit from progress in modern software engineering and if it is to have access to a software components industry. Although Ada satisfies these requirements, further experimentation will indicate whether it is a proper choice for manufacturing.

The language translation that we have developed is a pretranslator whose input is the entire Ada program and whose output is a collection of Ada programs, one for each target processor. Each of these in turn is compiled for its target processor using the pertinent compiler. A major function of the pretranslator is to insert Ada statements that handle interprocessor communication. As indicated in Figure 4, the part of the robot's internals shown in the system control computer could be the communication software inserted in this way.

Formal Semantic Models

Turning now to formal semantic models, the obvious questions are (1) what is being modeled and (2) why are the models needed? The logical views of such things as cells, factory floors, and process plans are being modeled, and these models are needed to develop the real-time control algorithms. Each manufacturing situation involves what is essentially a real-time—probably stochastic—scheduling problem. To control a manufacturing situation, it must be described and a scheduling algorithm developed. In addition, models are needed as a tool to treat exceptional conditions in an orderly manner and as a foundation for creating generic software components.

This can be best illustrated by again referring to Figure 3. To achieve a formal semantic model for the logical view of the component cell, it is necessary to have models for all the components shown in the figure. In addition, there must be an orderly means of combining these models as new components are created by assembling other components. For example, the model of cell will be some combination of the models of machines and transport, and the model of transport, in turn, will be some combination of the models of robot and vehicle. Use

of the phrase "some combination" is meant to emphasize that merely knowing the models for the logical view of the individual devices—that is, for the NC machines, the robot, and the vehicle—is not enough. The individual models obviously do not encompass information on the logical aspects of the mechanical interactions of the devices and the constraints of the factory floor. For example, the model of the robot says absolutely nothing about the vehicle. In addition to mechanical interactions, an understanding must exist of the way the devices are interconnected by the software in machines, transport, and cell. In other words, the approach to software components and the needs for models are entangled with each other.

The formal modeling system (Naylor and Maletz, 1986) employed here grows naturally out of the logical view. The state of the model is, roughly speaking, all the logical conditions that are true at a given moment, and state transitions are caused by the execution of "changes," where a change causes logical conditions to be altered. "Changes" are similar to rules in rule-based systems. They differ from rules in that their execution typically has an associated delay (for example, moving a vehicle from A to B takes time), and changes may occur simultaneously because the factory floor is a highly parallel system.

Process plans are another source of entanglement of software and formal models. In a sense, process plans can be thought of as a kind of program; the factory floor can be viewed as a kind of computer on which these programs execute; and the real-time control software can be thought of as an operating system that ensures the efficient execution of process plans. Scheduling processes in the manufacturing system is far more complicated, however, than scheduling program elements in a computer operating system. In computer operating systems so little is known about future jobs that scheduling has to be simple, and some version of a "fairness" criterion of efficiency is usually the only practical alternative. By contrast, in a manufacturing system a job can be considered as a part going through a step in a process plan. Thus, a great deal can be said about future jobs merely by knowing where each part is in its process plan. How to use this additional information is both a complication and a challenge. One is attempting to orchestrate the physical movement of parts on the factory floor simultaneously with their movement through the appropriate process plan. Constraints on both of these environments make the scheduling difficult.

There are two important things to be said about this approach to process plans. First, process plans are treated as separate and identifiable. That is, the description of a process plan is never to be intermingled with the description of the factory floor. In fact, a process plan is incorporated as a software component in much the same way that, say, a robot is. Figure 5 shows two process plan components and a factory floor component. Each process plan component contains information about steps in the plan and the ordering of these steps. The three components are assembled into the Manufacturing System component. The importance of the sharp separation between process plans and the factory floor is the reusability of each component. The factory

floor component can be used with other process plans, and, where meaningful, a process plan component can be used with other factory floors. It is clear that allowing the process plans and factory floor descriptions to be intermingled would severely limit such flexibility.

The other thing to be said about this approach is that the process plans are modeled formally in exactly the same way as any other component. The state of the model shows the parts that are currently progressing through the process plan and where each part is in the plan. Changes show what process plan steps are allowed next for a given part. In effect, this is a model of a logical view of the process plan. This means, among other things, that models of the process plans and the factory floor can be combined into a model that simultaneously exhibits the constraints of the factory floor, the constraints of the process plans, and the interconnections among these constraints. In other words, these components are assembled in basically the same way that any components are assembled.

Of course, each step in a process plan is not an atomic action, as has been implied by the discussion to this point. However, most of the details of a step can be suppressed when one is concerned with controlling the logical view of a manufacturing system. For example, a step might be the execution of a parts program. In the logical view this execution would usually simply be a change that takes a certain amount of time. The individual statements in the parts program would be irrelevant.

Generic Software Components

The research approach proposed here is a coordinated blending of software components, a common distributed language, and formal semantic models. Although we believe that this approach alone will have a major impact on software costs and flexibility, there is more that can be done. In par-

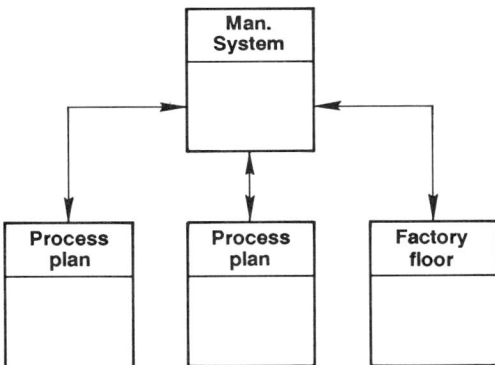

FIGURE 5   Assemblage of process plans component and factory floor component.

ticular, "generic software components" can significantly improve software reusability. This phrase is used far more generally than computer language specialists normally use it. Indeed, some might take exception to this usage. Still, for present purposes a generic software component will be something to which one adds information and, perhaps, other components to obtain a software component. Rather than becoming entangled in a definition, let us present two examples—one fairly simple and the other quite ambitious.

Consider a family of similar material transport systems. They all use the same kind of vehicle, but the number of vehicles may differ from system to system. Although each system has a "road system" or geographic layout, the road systems for two material transport systems may differ. As shown in Figure 6, any one of these material transport systems is a software component that is, in turn, an assemblage of vehicle software components and a road system software component. Although different from system to system, it should be obvious that the software that implements these various assemblages can have much in common. Each is doing essentially the same thing with similar software components. The goal of a generic component in this case would be to write one piece of software that would work for any one of the material transport systems. It would be supplied with information about the road system and

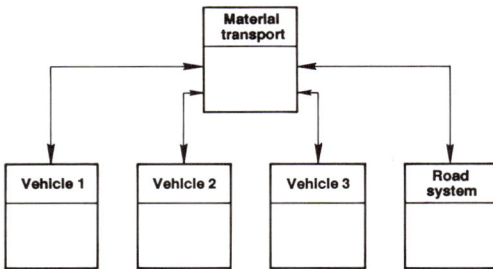

FIGURE 6   One of a family of material transport systems.

the number of vehicles. Further, it would be supplied with components to be assembled—that is, the vehicles and the road system.

The key point, then, is that one piece of software can be used, even "reused," over the entire family of material transport systems. Moreover, in this simple example it seems practicable and it is not difficult to imagine similar cases in which such generic software components would work. From the earlier discussion of a software components industry, one can also imagine buying such generic components.

Next, consider a far more ambitious case, one involving a family of factory floor and process plans. In a manner similar to the previous example, one would like to have a generic software component that would, in effect, be a completely flexible factory controller. The concept is as follows: The generic component would be given descriptions of a particular factory floor, a particular set of process plans, and a description of the interconnections between them; then, based on these descriptions, a real-time controller would be realized without the need for reprogramming. This is obviously a very ambitious goal, one that is probably not completely attainable. In a sense, however, it constitutes an upper limit for generic components. It is also an instance of generics touching the world of artificial intelligence (AI), since, to the extent that such a controller would be realizable, it would undoubtedly rely on AI techniques.

Finally, given this concept of generic software components, an obvious issue is the form of the information given to the controller. The approach proposed here is to use formal models. For example, the information given to the generic factory controller would be a formal model of the assemblage of the factory floor and process plans. This, then, is the other use of formal models and another blending of the various parts of this research approach.

AN APPLICATION

Let us now consider an experiment that applies this approach. This is an experiment with a partially generic factory floor controller. Figure 7 gives an overview of the experiment. A real-time simulator of a factory floor is contained on one VAX computer, and the generic factory floor controller is on another VAX. The common distributed language is the distributed version of Ada. Software components are Ada packages and tasks.

An advantage of treating the factory floor as an assemblage of software components, for which we have formal semantic models, is that simulation is extremely easy. One merely replaces the private internals of lower-level components and leaves the public interfaces and their interconnections unchanged. "Lower-level component" means device components such as robots and NC machines. The replacements are easily selected so that the real factory floor and the simulation have the same formal model. The simulator on VAX 1 is created in this way. As far as the controller on VAX 2 is concerned, VAX 1 behaves exactly as the real factory floor would.

Now consider the factory floor controller on VAX 2. It contains another simulator of the factory floor, the tracking model, whose function is to keep track of the current state of the factory floor, that is, the current state of the simulator on VAX 1. The controller also contains a simulator, the search model, of the combination of the factory floor and process plans. It is used by the controller to explore alternative future scenarios, and it uses these explorations to select the next commands to send to the factory floor on VAX 1. The tracking model and the search

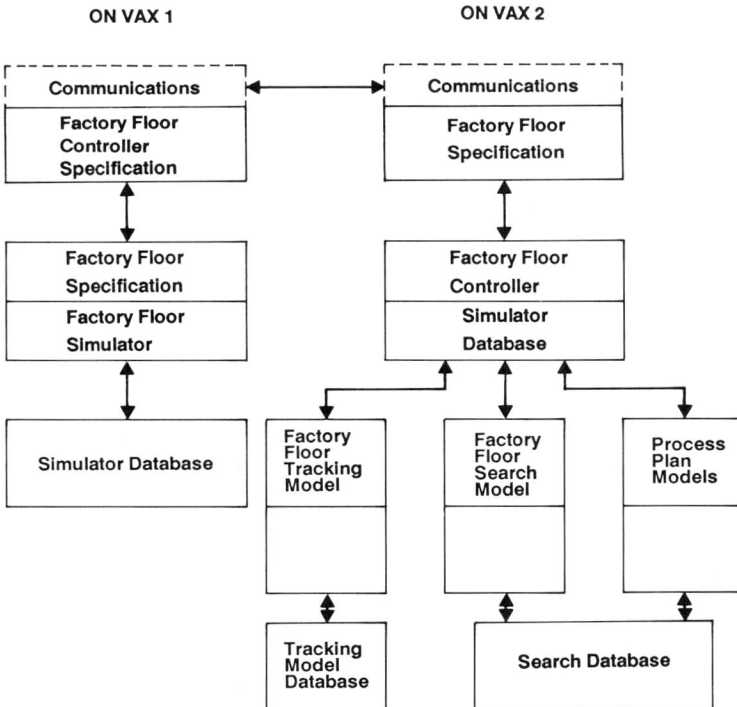

FIGURE 7 Factory floor simulator and controller.

model are created in essentially the same way as the factory floor simulator on VAX 1. Each is an assemblage of software components. This is also true of the factory floor part of the search model. In other words, the same assemblage of software components and semantics is used in three different places in this application.

Since the purpose was to experiment with software components, their assemblages, and formal models, the controller is quite simple. After each change in the state of the factory floor, the controller explores different future scenarios using the search model, starting from the state contained in the tracking model. A few simple pruning rules keep the search manageable; clearly, more could be done. In a sense, this application creates a test bed within which various control algorithms can easily be examined. In any event, the controller is partially generic in the sense that it will work with almost any tracking model and associated search model. That is, the two models are the information supplied to the generic controller.

Figure 7 also shows a typical result of using the distributed language pretranslator. The two dashed boxes labeled "Communications" would be inserted by the pretranslator.

RESEARCH QUESTIONS

We have stated our basic assumptions about how control software should be designed and developed, and we have discussed each of them. However, this is only a beginning. It is certainly not a detailed recipe for software development, nor can one currently purchase control software developed following this prescription. It will take about 10 years to realize this vision, particularly the generic part of it. Ten years may sound pessimistic, but considering how long the Ada and MAP efforts are taking,

10 years appears to be realistic. After all, this is a far more complex problem than MAP.

It is relatively easy to see the major research directions that should be followed initially. Although the fundamental ideas of software components are known, much remains to be done in the area of distributed languages to support them, particularly considering real-time issues and the fact that one is almost certainly dealing with heterogeneous systems.

There are also many important issues at the boundaries between the basic assumptions contained in this proposal. One is the boundary between software components and formal semantic models. We need a better understanding of this interplay. Formulating a model for a factory floor is, among other things, an awesome task; therefore, assembling a model from models of software components as the components are assembled—perhaps with computer aids—becomes extremely attractive. We also need to extend those ideas to the assembly of generic software components. Furthermore, the development of control algorithms—eventually generic ones—should take advantage of the system description as an assemblage of smaller descriptions. In particular, the application of artificial intelligence techniques to such assemblies must be explored.

At the boundary between software components and distributed languages, the basic question is: Can all this be done without a crippling sacrifice of real-time performance? One must be concerned whether the assemblage-of-software-components structure and the relegation of this structure to the distributed language translation system may combine to produce unacceptable delays. Another issue concerns distribution, specifically, how the distribution of a software component over several processors should be evidenced in the assembly of software components and their models.

SUMMARY

We have argued that research on integrated manufacturing systems should be integrated in the sense that a number of interrelated problems should be attacked together in a unified manner. We have also admitted that we do not know how to do this in general. However, we have been able to make some progress on a particular problem, and we believe that we can see how to carry out the required further research in a unified manner.

REFERENCES

Kaminski, M. A. 1986. Protocols for communicating in the factory. IEEE Spectrum 23(4):56–62.

Naylor, A. W., and M. C. Maletz. 1986. The manufacturing game: A formal approach to manufacturing software. IEEE Transactions on Systems, Man, and Cybernetics SMC-16:321–334.

Rentsch, T. 1982. Object oriented programming. Sigplan Notices 1(9):51–57.

Shaw, M. 1980. The impact of abstraction concerns on modular programming languages. Proceedings, IEEE 68(9):1119–1130.

Volz, R. A., and T. N. Mudge. 1984. Robots are (nothing more than) abstract data types. In Proceedings of the SME Conference on Robotics Research: The Next Five Years and Beyond. Robotics International. Dearborn, Mich.: Society of Manufacturing Engineers.

Volz, R. A., and T. N. Mudge. 1987. Timing issues in the distributed execution of Ada programs. IEEE Transactions on Computers, Special Issue on Parallel and Distributed Processing, C-36(4):449–459.

Volz, R. A., and A. W. Naylor. 1986. Final Report of the NSF Workshop on Manufacturing Systems Integration. Technical Report RSD-TR-17-86. Robotic Systems Division, Center for Research on Integrated Manufacturing, College of Engineering, University of Michigan, Ann Arbor.

Volz, R. A., T. N. Mudge, A. W. Naylor, and J. H. Mayer. 1985. Some problems in distributing real-time Ada programs across machines. Pp. 71–84 in Ada in Use: Proceedings of the 1985 International Ada Conference. New York: Cambridge University Press.

Volz, R. A., T. N. Mudge, A. W. Naylor, and B. Brosgol. 1986. Ada in a manufacturing environment. Pp. 433–440 in Proceedings of the Fifth Annual Control Engineering Conference, Rosemont, Ill.

Volz, R. A., P. Krishnan, and R. Thierault. 1987. An approach to distributed execution of Ada programs. Pp. 187–197 in Proceedings of the Workshop on Space Telerobotics. Pasadena, Calif.: Jet Propulsion Laboratory.

Volz, R. A., T. N. Mudge, G. D. Buzzard, and P. Krishnan. In press. Translation and execution of distributed Ada programs: Is it still Ada? IEEE Transactions on Software, Special Issue on Ada.

PROCESS AND ECONOMIC MODELS FOR MANUFACTURING OPERATIONS

Vijay A. Tipnis

ABSTRACT   Unit manufacturing operations are at the heart of every manufacturing system. Process and economic models are essential tools in designing, developing, planning, optimizing, and controlling manufacturing operations and systems. The status of the development and application of these models is presented along with the real-world problems and challenges that influence their use.

INTRODUCTION

At the heart of every discrete parts manufacturing system, whether traditional or flexible, attended or unattended, are manufacturing processes that convert input material into a prescribed part or assembly configuration. The central purpose of every manufacturing system is to achieve the transformation at the most desired production rate and cost. All other operations, including data flow, material handling, set-ups, loading and unloading operations, inspection and quality control, preprocessing, resource supply, and support systems such as tooling, maintenance, and cleanup, must be considered to be in support of the transformation of a starting material into a final product.

Most of the improvements that have resulted from automated manufacturing systems can be identified with a drastic reduc-

tion in the time and cost associated with these nonprocessing support operations. The challenge for future automated systems is to continue to accomplish a reduction in these nonprocessing operations while also encouraging unattended operation for a predetermined period of time. Achievement of this will require a sufficient understanding of the processes to allow the construction of reliable models for designing, planning, optimizing, and controlling unit manufacturing processes.

Modern manufacturing system design is still evolving into a cohesive methodology where diverse technologies of design, material science, material processing, numerical control, quality control, material handling, sensors, computer networks, computer software, data-base systems, and man-machine interaction must be integrated. The role of processing is crucial in accomplishing this.

Economic models, as complements to

process models, have evolved for unit manufacturing processes, sequences of manufacturing processes, and total manufacturing systems. Although progress to date shows promising applications, further research is needed to accommodate various materials, process types, and complete manufacturing systems in these models. Furthermore, the economics of product and process design and development must be better understood to ensure that the designs are producible at the desired cost level.

Some of the crucial gaps that still exist in the design-manufacturing interface and some of the deficiencies in the state-of-the-art process and economic models are outlined in this paper. In addition, an attempt is made to relate the actual manufacturing process to the manufacturing system design.

ACHIEVING THE DESIGN INTENT

The design intent can be achieved in a variety of ways. Material type, part features, tolerances, finishes, and fit requirements can often be modified without jeopardizing the part, assembly, or component function. Such modifications in design can ensure that a cost-effective manufacturing process will be used. This has been demonstrated for discrete parts used in the aerospace, automotive, and precision parts industries, where a large fraction of the costs can be influenced through an effective design-manufacturing interface.

Although a formal organization dealing with the design-manufacturing interface does not exist in most corporations, it is not uncommon for an ad hoc personal relationship to exist between the design and advanced manufacturing groups. While this arrangement can deal with important aspects of an issue, the focus is often limited to specific product items. A stronger interaction and a formal communication link between design and manufacturing are clearly needed and can contribute to achieving a

design based on the requirements for assembly, service, and maintenance throughout the life of the product. Furthermore, this should ensure that the design will be rationalized for the capabilities of the manufacturing system that is going to convert the design intent into reality. To accomplish this will require that some significant unresolved issues be addressed. Several of these are discussed in the following paragraphs. Subsequent sections discuss a number of them more extensively.

Representation of the Physical Object

It is well known that engineering drawings (views or isometrics) do not guarantee that the object represented is physically realizable. Imaginary objects can be represented on paper or on a computer graphic system. Although computer graphics has progressed through several stages of representations, including wire frames, polygon schemes, sculptured surfaces, and solid modeling, there are no intrinsic criteria to assure that a drawing represents a physically realizable object.

The problem of guaranteeing a physically feasible object requires (a) a validation for checking internal consistency of the physical features of the object, (b) a criterion for ensuring against under- or overdimensioning, and (c) a consistency and adequacy test for tolerances. While currently available solid modeling systems address some of these problems, more work is needed to establish validation criteria based on the topology of the objects. Over- and underdimensioning have been the subject of research for the past decade (Hillyard and Braid, 1977; Requicha, 1977; Light, 1979). While constraints on geometry can indicate whether the drawing is under- or overdimensioned, there are no available criteria to determine which of the dimensions are under or over. In spite of these deficiencies, interactive conceptual design and drawing systems using computers have been commercialized by Metagraphics Company and by Cognition Company. Until these limitations are eliminated, the manufacturing engineer must continue to decide if the representation is physically realizable, whether it is over- or under-dimensioned, and whether the specified tolerances are consistent.

Tolerancing of the Drawing

Tolerancing is a convention that started during the late 1920s and had matured to the ANSI Y 14.5 standard by 1973 (Voelcker, 1988, in this volume). No existing mathematical theory ensures uniqueness, consistency, or completeness for the tolerances of a drawing. Current computer graphic systems, whether wire frames, bounded surfaces, sculptured surfaces, or solid models, present nominal dimensions. Tolerances are merely attached as labels. Although IGES/PDES committees are working to develop such capabilities, most computer data bases cannot currently capture tolerances. A draft of PDES (Smith, 1987) gives representation formats for geometry and tolerancing. Surface finish and surface integrity remain to be addressed.

Furthermore, a toleranced drawing does not represent a unique physical object. Since different manufacturing processes can generate objects with distinct tolerance bands and distributions (Bjorke, 1978), it is not uncommon to find that parts manufactured to tolerance limits may not be capable of assembly (Whitney et al., 1988, in this volume). Selective assembly, part mating, and tolerance stacking are often used as a means of compensating for these inadequacies.

Process Determination

From a dimensioned and toleranced drawing, and specifications for the material and a determination of the application constraints, the next step is to derive a complete set of process sequences for production of the physical object. Not only are there no criteria and methodologies to determine these steps automatically, the procedures used by highly skilled and experienced manufacturing engineers do not yield unique answers. Manufacturing engineers commonly begin by comparing the part size, shape, features, material, and tolerances against the process capabilities. They then proceed (a) from the goal of the specified finished object to intermediate steps by adding the necessary "stock allowance" at each preceding processing step or (b) from a target blank to the specified finished object by subtracting the allowance or (c) by applying both (a) and (b) schemes alternately. The logical representation of the selection procedure cannot be readily characterized.

It is clear, however, that each process is capable of generating a specific tolerance distribution on a given material, and hence, regardless of the designer's specification, the selected process and material combination truly determines the tolerances on the manufactured object.

Trade-Offs Among Features, Tolerances, Quality, and Cost

Achieving an optimal design requires careful consideration of all aspects of the product and the manufacturing system. The ad hoc interaction between design engineering and manufacturing engineering frequently occurs as shown in Figure 1 (Tipnis et al., 1978). The part design concept goes through a series of iterations in which design considerations of function are weighed against manufacturing considerations of productivity and cost within the context of the prescribed quality levels. The manufacturing engineer determines the possible trade-offs between cost and such attributes as features, tolerances, and quality. The challenge is to formalize the interaction so as to ensure that a complete set of trade-offs has been derived.

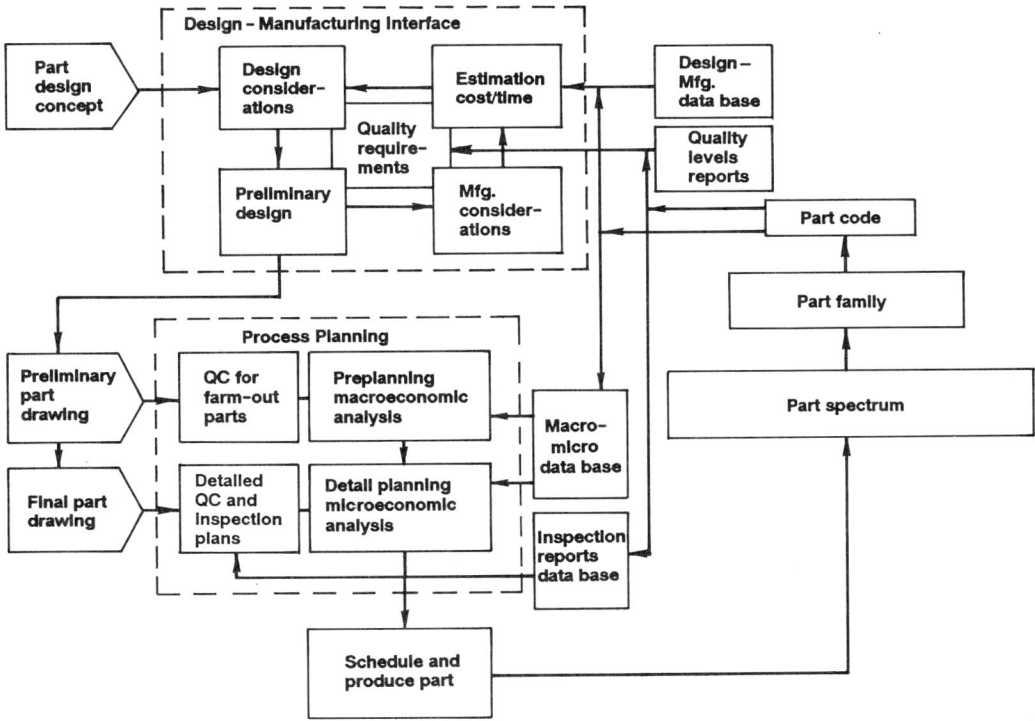

FIGURE 1   Representation of design-manufacturing interface in the product-manufacturing system.

This issue is being addressed in a variety of ways. In the aerospace industry, the practice of constructing life-cycle cost models to evaluate alternative conceptual designs is known to produce significant reductions in cost (Shoemaker, 1980). This practice allows all aspects of the design to be treated as a system, including items such as product and process R&D, acquisition, support, and fuel use. As can be seen from Figures 2 and 3, the relative impact of an improvement in the process can be weighed against the overall cost. A similar life-cycle view is becoming popular in the automotive industry (Compton and Gjostein, 1986).

The foregoing discussion suggests that drastic cost reductions should be achievable if the design and manufacturing group is allowed more freedom in the interpretation of the design intent and can evolve, therefore, significant modifications while main-taining the original design intent. In this way, some of the disadvantageous effects and tunnel visions arising from an early crystallization of the design can be overcome. The usual rather narrow path followed during detailing of the design of components and parts and their assembly suggests that the creative process of design synthesis and design analysis needs to be better understood.

There is no doubt that considerations of manufacturing, assembly, serviceability, maintainability, and use of the part or component during its entire life cycle would benefit from intensive functional and cost-effective designs. Organizations that promote such interaction are known to produce outstanding products (Whitney et al., 1988, in this volume). Why some organizations are able to do this well needs to be better understood. The construction of a

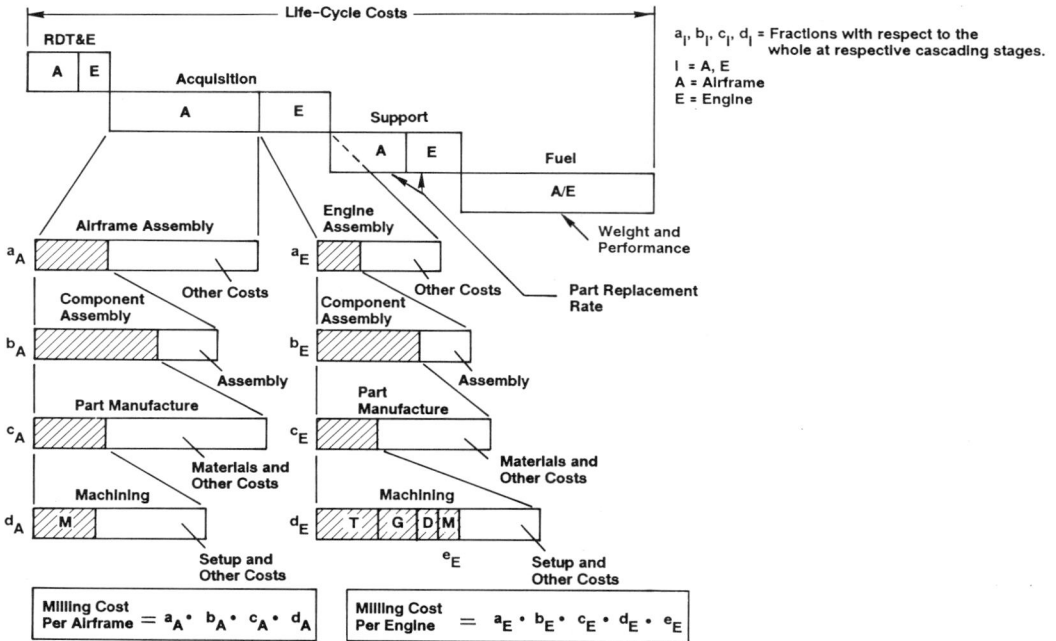

FIGURE 2   Economic opportunity windows during conceptual feasibility testing of a new material removal technology.

model and a methodology for describing the life cycle of a part should be the focus of serious investigations.

It can be demonstrated that the design intent should be weighed against manufacturing process realities only within the context of the overall mission and the life-cycle costs. Whether the existing manufacturing system constraints should dictate the part design depends on whether the mission requires new materials and therefore new or improved processes. The concept of "flexibility" of a manufacturing system (Tipnis and Misal, 1985) has become a key element in establishing the degree of freedom that design engineers should be allowed for parts to be cost-effectively manufactured in the system. Thus, the interface between the product design and modern manufacturing processing has become tightly coupled. This area deserves a rigorous investigation.

PHYSICAL PROCESSES IN MANUFACTURING

From ancient times, implements have been shaped from materials such as wood, clay, sand, fiber, and stone by using tools and processes that have been developed largely by trial and error. These shaping processes were clearly the forerunner of the historical development of manufacturing processes driven by the impetus to improve naturally occurring materials through mining and winning, refining, alloying, and other methods. Current manufacturing processes, which are limited to about 100, can be grouped, as in Table 1, according to the physical processes used to convert the input raw material into the prescribed configuration (part of assembly).

Most manufacturing processes are in reality a series of individual (unit) processes through which the input material is "processed" until a prescribed configuration is achieved. Each of these unit processes involves a series of steps in which material conversion occurs and various supporting activities take place, including, for example, the positioning of the workpiece, adjustment of the tooling, or inspection of the part. A proper description of this collection of operations, often referred to as a processing sequence, requires an understanding of the technologies involved in each of the units. As mentioned earlier, no unique sequence of processes can be assumed to exist for creating a part or component. For example, an automotive connecting rod can be manufactured by a variety of operations

from a cast, forged, powder metal compact, or near-net-shape forging or casting.

New and improved materials have created a demand for new and improved manufacturing processes. Many applications now demand that materials perform at increasingly high temperatures and high strength levels (Clark and Flemings, 1986). New applications have created a demand for material processing methods that can shape objects of complex configuration, accurate dimensions, and tight tolerances. Some materials are now being processed in a fashion that leads to properties near their theoretical limits. Ingenious combinations of microscopic structure, alloying, reinforcing, coating, deposition, and other tech-

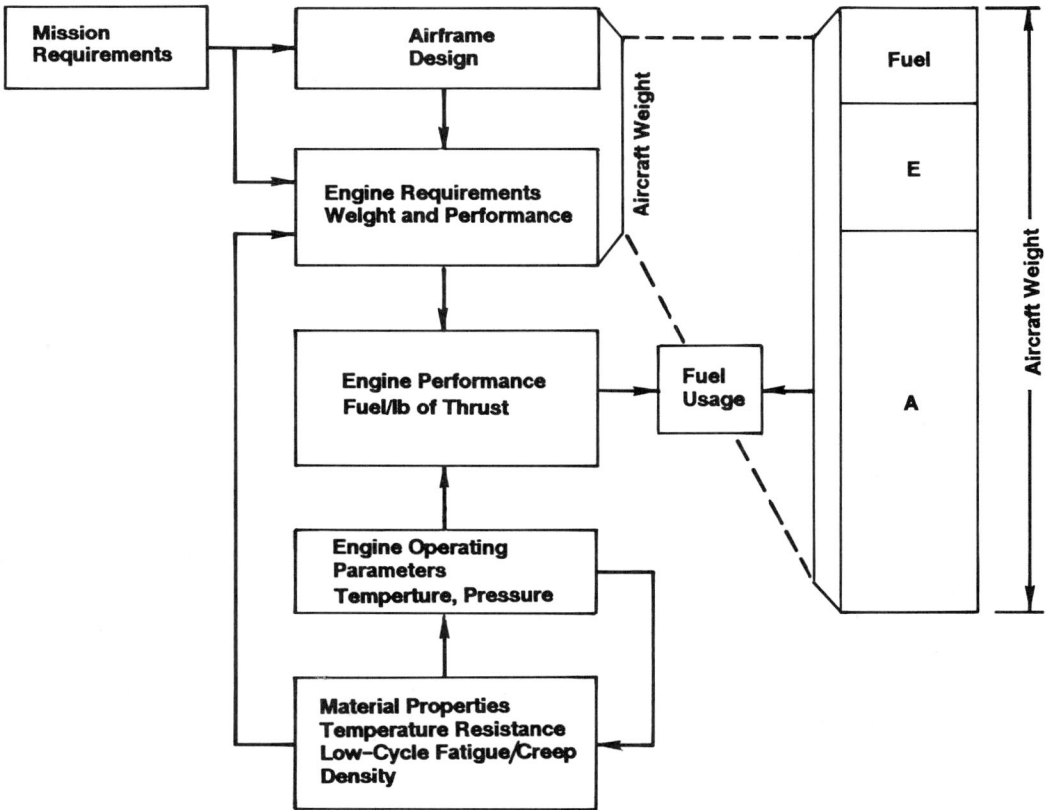

FIGURE 3   Influence of engineer performance and weight on fuel-use costs.

TABLE 1   Traditional and Nontraditional Manufacturing
Processes (a representative list)

Process	Traditional	Nontraditional
Material Removal		
Machining	Turn, mill, drill, bore	EDM, ECM, laser, EBM
Grinding	ID/OD, surface, belt	ECG, EDG, creep-feed
Punching	Shearing, stamping	Laser cutting, plasma
Material Deformation		
Rolling	Plane, rounds, tubes	
Forging	Open-die, closed-die	Isothermal
Extrusion	Forward, backward	
Wire drawing	Open-die	Hydrostatic
Forming	Sheet, tube, spinning	Superplastic, explosive
Material Addition		
Plating	Electro, electroless	
Coating	Thermal spray	Chemical vapor deposition, powder vapor deposition, sputtering
Material Joining		
Welding	Electrode, plasma arc	Laser, electron, inertia
Brazing	Thermal	Electrothermal
Compaction	Puddling	Powder metal
Material Transformation		
Solidification	Casting, lost wax	Directional, continuous
Heat treatment	Tempering, annealing	Laser beam heat treatment
Alloying	Deoxidation, melt	High pressure

niques are deployed to "design" a material for a specific application, often with a unique combination of end-use requirements for strength, ductility, wear resistance, working temperature range, corrosion resistance, and so on. Besides the traditional use of mechanical and thermal energies, many of the new manufacturing processes use chemical, electrical, magnetic, laser, electron beam, plasma, or combinations of two or more of these energy sources.

It is important to recognize, however, that a new manufacturing process rarely displaces a traditional process completely. Instead, each new process tends to fulfill a special need where it is superior in performance and cost-effectiveness to all other alternatives. As new processes have evolved

to meet special needs, traditional processes have undergone continued improvements in response to the new materials and the demand for higher performance. Thus, it is no surprise that traditional manufacturing processes continue to play a major role in manufacturing.

It is increasingly important, therefore, that material processing techniques, whether new or traditional, ensure that (a) the resulting product has the desired end-use properties, (b) the process rate is acceptable for the production requirements, and (c) the total cost including material and processing is economically justifiable in relation to other alternatives. An understanding of the applicability, capability, and processability of each new and traditional method is essential to the productivity, quality control, and economics of manufacturing systems.

Process development, involving the translation of the laboratory research on process design into a full-scale production process, has traditionally evolved along an experience learning curve. Consequently, costly trial-and-error procedures are frequently repeated. Few academic researchers have been attracted to investigating the technological and economic problems of production scale-ups of discrete parts manufacturing processes. An increased interest in and attention to these problems is clearly warranted.

MODELS OF PHYSICAL PROCESSES

An improved understanding of the capabilities, constraints, and limits of processes can be of significant importance in improving the quality, productivity, and cost reductions of manufactured parts. Before the 1850s, process knowledge resided within the expertise of artisans. Little formal documentation existed for this information. More recently, attempts have been made to understand and document physical phenomena that involve processes. These efforts have generally been of two forms: phenomenological investigations aimed at basic understanding of the process and empirical investigations aimed at determining the best operating conditions for a given process.

Process Knowledge

Despite the progress on methodologies for process modeling, most process knowledge remains locked in the expertise of a few individuals associated with the process. In many cases there is little phenomenological understanding of the process. The real challenge is how to extract this knowledge and reconcile it with phenomenological and empirical insights. What does not work is often more useful than what works. Until

the advent of expert system methodologies, no systematic approach was available. This know-how consists of

- Rules of thumb learned from experts, from peers, and by trial and error;
- "If A then B" rules or knowledge or alternative frames of reference; and
- Observations of catastrophic failures and limits of a process.

Process knowledge extraction and presentation for real processes require a close partnership between an expert practitioner and a process researcher experienced in knowledge engineering.

The potential benefits of such models in designing, developing, planning, optimizing, and controlling the manufacturing processes are great and are the basis for much of the discussion in subsequent sections of this paper.

Phenomenological Process Models

Phenomenological models are constructed to describe the cause-effect relationships between the basic input variables and the output variables of a manufacturing process. The drive toward creating phenomenological models is a natural extension of the belief that, since we understand the basic laws of physics, it should be possible to apply these laws and define manufacturing processes mathematically. Although this has been a desirable goal, there are some formidable difficulties that have prevented the development of practical phenomenological process models for manufacturing (Ford, 1966; Shaw, 1966; Opitz, 1966). The following generalizations can be made about the current status of these models:

- The available theories of plasticity, friction, wear, instability, fracture, and catastrophic failure do not readily apply to the extreme ranges of stresses, strains, strain rates, temperatures, and pressures within the working zone of most processes.

• The implicit assumption that the material is continuous does not conform to the properties of real materials, which are non-isotropic and contain nonuniform distributions of inclusions, voids, and multiphases. Minute changes in the composition and microstructure of a material may induce a profound change in its processability.

• Most processes are non-steady-state and cannot be treated by the usual steady-state techniques.

• Processes are time-varying in that they tend to degrade from self-induced and external disturbances, such as vibrations, friction, wear, plastic flow, and kinematic instabilities.

Although phenomenological models are not yet sufficiently refined for many practical applications, the insights gained through their investigation have proved valuable for achieving process improvements. They can often reveal crucial characteristics that will make new process development much easier or will lead to significant improvements in existing processes.

Empirical Process Models

Empirical models relate process performance directly to process variables using experimental data from a real process or a closely simulated situation. As shown in Figure 4, the empirical model can be viewed to encompass one or more phenomenological models dealing with the specific cause-effect relationships.

The approach often followed in creating a process model includes the following steps (Tipnis, 1977a):

• Observe a real process or its effects.
• Simulate the real process under controlled conditions.
• Establish cause-effect relationships, if feasible, between the basic process variables and their results using physical laws.
• Establish direct empirical relationships between the process variables and process performance.
• Predict what will happen in the real process.
• Improve the correspondence between the model predictions and the real process.

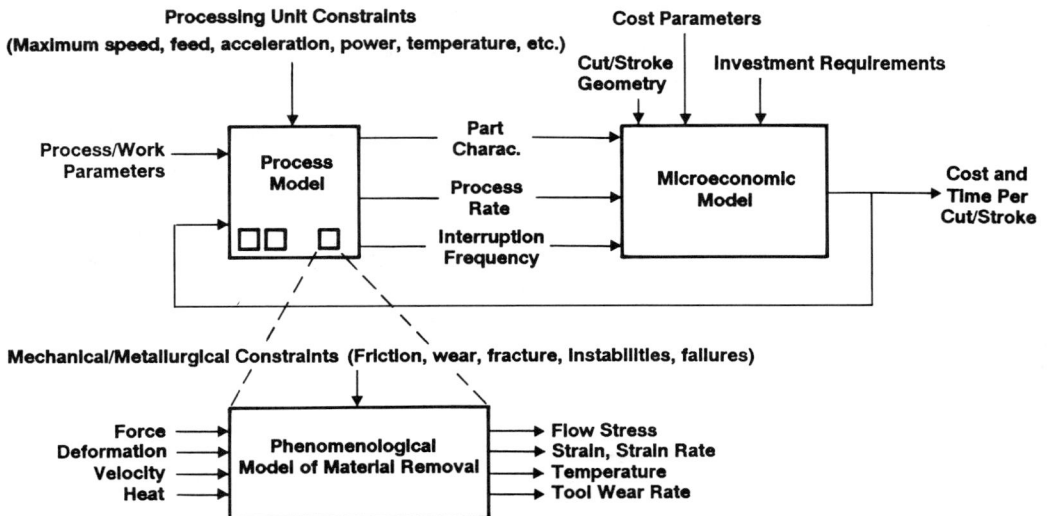

FIGURE 4   Relationship between process and microeconomic models.

For construction of empirical models, statistically planned experiments and response surface methodologies have been successfully applied (Wu and Ermer, 1966). The status of empirical models is roughly as follows:

• The usefulness of an empirical model is critically dependent on the design of the experiments, the selection of the model, and the variance.

• An appropriate working region for the model must be carefully defined, since empirical models of objective functions as well as constraints must be constructed.

• Most empirical models are first- or second-order equations in log-transformed space, making them useful for controlling a process through software.

• Practical implementation of empirical models, although feasible, is yet to be fully exploited.

Recently, time series analysis has also been applied with some success to a few machining processes. This approach has not yet made a significant impact on process modeling (Kapoor and Wu, 1980).

PROCESS ECONOMICS

The goal of economic models is to create a tool that can be used to determine a set of operating conditions that will optimize the economic objective function within the working region of the process. Process economics has been investigated since about 1900 (Taylor, 1907). At the micro level, process economics deals with optimization and control of unit manufacturing processes. At the macro level, it deals with sequences of processing units and support systems and can be used to investigate the economics of the entire manufacturing system.

In considering the economics of a specific unit manufacturing process, it is necessary to define a characteristic processing rate and resetting frequency for the work material

and the set of operating conditions. Since a typical process is capable of producing configurations from a given work material that may affect the properties of the finished part or assembly (for example, a tolerance range, distribution, surface finish, and integrity), the prescribed performance requirements impose limits or constraints on the process variables. Also, constraints must be introduced to deal with regions where catastrophic failure in the process may occur—for example, tool chatter or instabilities that limit the range of operating variables. These constraints define the working region within which the process can be safely operated. Thus, the process rate and resetting frequency are functions of numerous operating variables that ultimately determine the production rate and the cost of the part or component.

Until recently, each process was described by a separate economic model. Optimization of the entire system involved tedious sequential differentiation of each variable taken one at a time. Unfortunately, this method does not guarantee that an optimum solution will be found. The following discussion of economic optimization draws heavily on applications to machining operations (Tipnis et al., 1981). Although similar models should be applicable to all manufacturing processes, this remains to be demonstrated.

Generalized Economic Objective Function

A general economic model for material removal processes is shown in Figure 5. The quantity I represents an objective function that can be transformed into time, cost, or a combination of time and cost (for example, profit rate) by an appropriate definition of the parameters. In this formulation the generalized process rate—that is, the material removal rate—and the generalized process resetting frequency functions—that is, the tool life—are treated as inde-

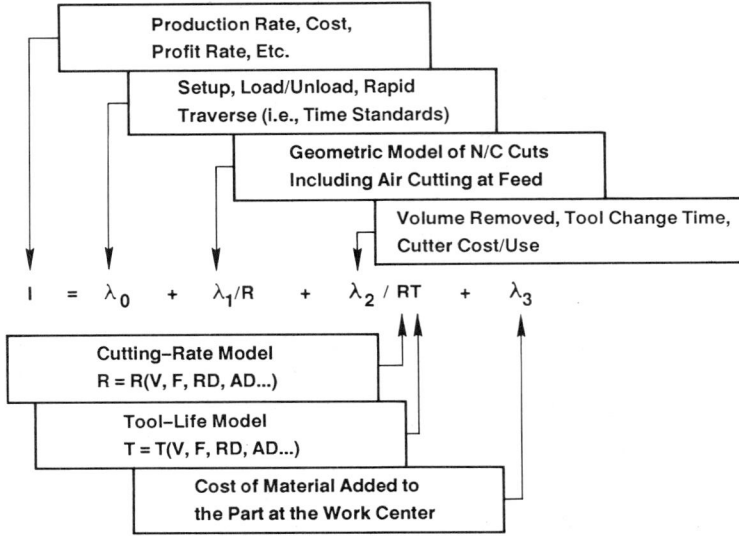

FIGURE 5   Generalized objective function for a process model of a material removal process.

pendent variables. However, material removal rate and tool life themselves are functions of operating variables such as speed, feed, and depth of cut. This leads to implicit dependence between the material removal rate and the tool life. Consequently, a trade-off function exists between the material removal rate and the tool life.

Constraints on the Objective Function

The range of allowable values for the variables must be constrained by the physically realizable working region. This places constraints on the maximum and minimum permissible values of operating variables, such as machine speed and feed range,

power, temperature, or pressure. Catastrophic failure limits from tool breakage and regions of nonpermissible vibrations arising from phenomena such as tool chatter also impose constraints. An example of the test data necessary to define cutter breakage constraint for end mills is shown in Figure 6. Empirical models for constraints can be derived using the same methodology as that for the process models.

Optimization Strategies

The most generalized optimization procedure involves minimization of the objective function, I, given in Figure 5, relative to either time or cost. In the case of machining, the minima for both time and cost have been discovered to exist for conditions that are described by the R-T-F curve (Ravignani et al., 1977). The equation of the R-T-F curve can be found by setting the Jacobian of (R, T) with respect to operating variables such as cutting speed, V, and feed, F, to zero; to ensure that the R-T-F curve represents a trade-off between the maxima of R and T functions, the Hessian should be negative. An example of an R-T-F curve is shown in Figure 7 for a milling operation. Similar curves can be developed experimentally for end milling, sawing, face milling, turning, and other machining operations.

If the values on the R-T-F curve are within the working region defined by the constraints, the process will be optimized if it operates along either the time or cost minima of the R-T-F curve. If the values on the R-T-F curve are beyond the working region, the optimal strategy would be to

operate at the limits of the working region closest to the R-T-F curve.

Furthermore, it has been demonstrated that the minimum processing time always occurs at higher material removal rates than the minimum for cost (Ravignani et al., 1977). The cost minimum is shown in Figure 8 on the plot of cost versus cutting rate. This approach is promising for other applications.

Control Strategies

A viable control strategy must maintain the processes within safe operating limits and operate at a prescribed economic optimum. The concept of a control strategy is illustrated in Figure 7. Since the R-T-F curve exists within the working region defined by the constraints, the correct control strategy should be to operate the process at or near the R-T-F curve. Operating close to

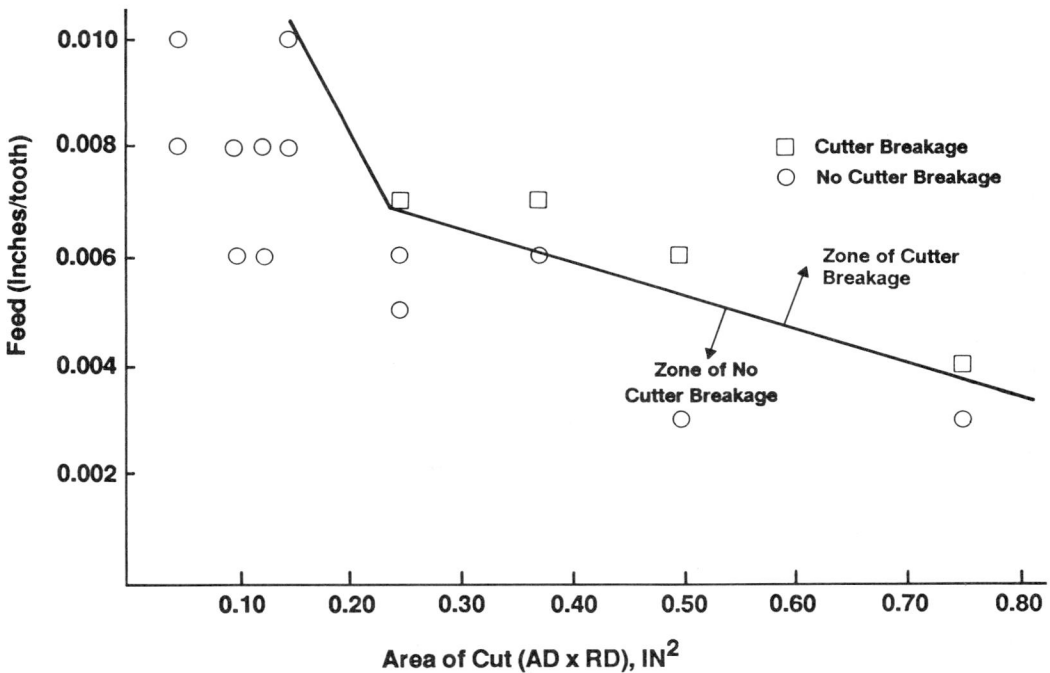

FIGURE 6   Cutter breakage constraint—feed rate versus area of cut for a 1-in. diameter, 2-in. flute length, M42 HSS end mill cutter.

4340 Steel, 217 BHN, Dia. = 1 in.
FL = 2 in.   4 Flute M 10 HSS Cutter,
AD = 1 in.   V = 150 fpm

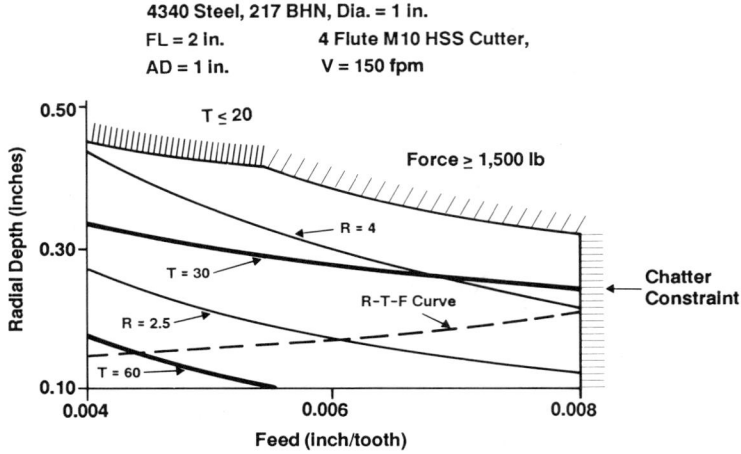

FIGURE 7   Working region defined by constraints and R-T-F for end milling operation; Time, T, in min; Rate, R, in in.^{3}/min.

the constraints involves risking catastrophic failures (Tipnis, 1977a, 1977b). Investigations aimed at determining working regions, trade-off functions, and control strategies for different processes that will allow safe operation near these limits should be pursued.

Traditionally, most discrete parts manufacturing processes have been controlled and guided by a machine operator. During the past two decades, a growing use of numerical control in machining operations has relieved the operator of the responsibility of guiding cutter motions during a cut. The operator is, however, still responsible for the quality of the parts produced and often inspects the parts after the process. Parts that fail to meet specifications because the process drifted out of control are either reworked or scrapped. Gradually, sensor-based process control has been applied to machining processes. As shown in Figure 9, direct size control can be achieved by a variety of methods, including manual (operator) compensation, post-process automated sensing and compensation, or in-process automated sensing and compensation.

The most promising process control is through on-line sensing that detects a drift or change in a dimension and makes the necessary compensation adaptively. This is particularly desirable for unattended machining operations.

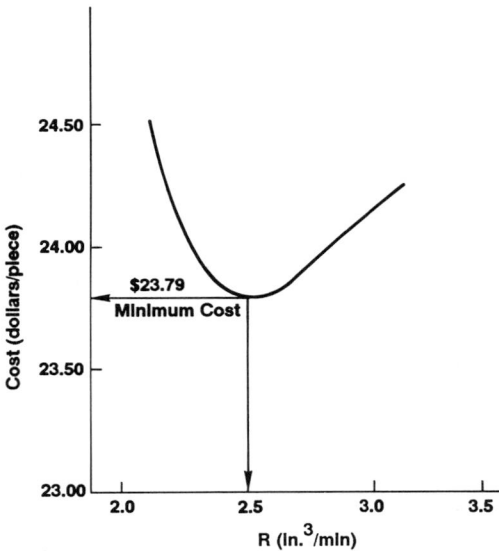

FIGURE 8   Cost versus cutting rate. Minimum cost occurs at 2.5 in.^{3}/min on the R-T-F curve (see Figure 7).

PROCESS DEVELOPMENT

The design of a process to achieve specific mission requirements has been well established within the aerospace industry for at least three decades. A mission requirement for a fighter plane to reach an altitude of 40,000 feet within 1.5 minutes directly translates into performance requirements for the engine, airframe structure, and avionics. These performance requirements can be translated into targets for working temperatures and pressures for the turbine and combustion chamber of the jet engine, for the structural strength and integrity of the airframe, and for the data and information needs of pilots and navigators. This development of material and structural requirements from the mission requirements has led to the practice of "designing" materials with the required combination of properties.

The concept and practice of creating a process to make the prescribed configurations from the "designed" material has evolved through interaction between materials scientists and manufacturing engineers. When new materials are involved, it is often discovered that the available processes are inappropriate or incapable of producing the prescribed configurations from the newly developed materials. Although the importance of close collaboration between materials and materials pro-

cess development has been well recognized within the aerospace industry, academic training of these two closely related groups still occurs in different disciplines. The materials, metallurgy, and ceramics departments continue to be separate from the materials engineering and mechanical engineering departments. A mechanism for close cooperation between industry and the university research community is needed to enhance designing of new processes.

Typical steps needed to translate mission requirements into detailed process plans are shown in Figure 10. Strategic activities dominate the process design and development process in the early phases. At the top of the figure are shown the different economic models that can be used in the development of the design and manufacturing process.

Development of New Processes

In this section, the issues and challenges in process design and development are addressed, with a primary focus on process and economic models. No attempt is made to cover other pertinent aspects, such as the role of materials development and broad manufacturing system issues.

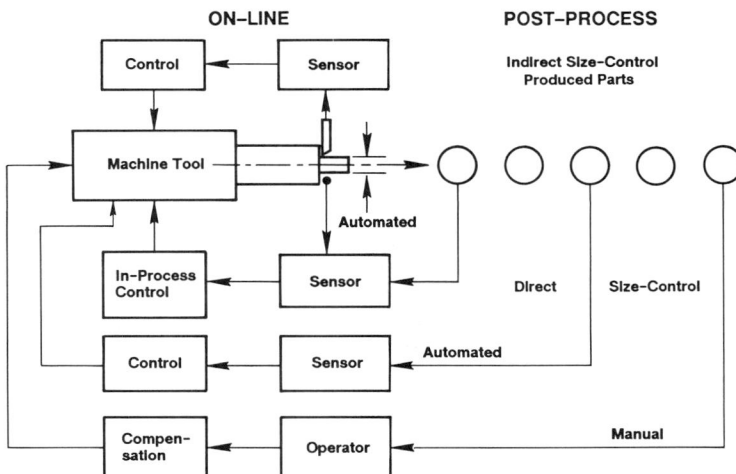

FIGURE 9   Different methods of size control (Novak, 1980).

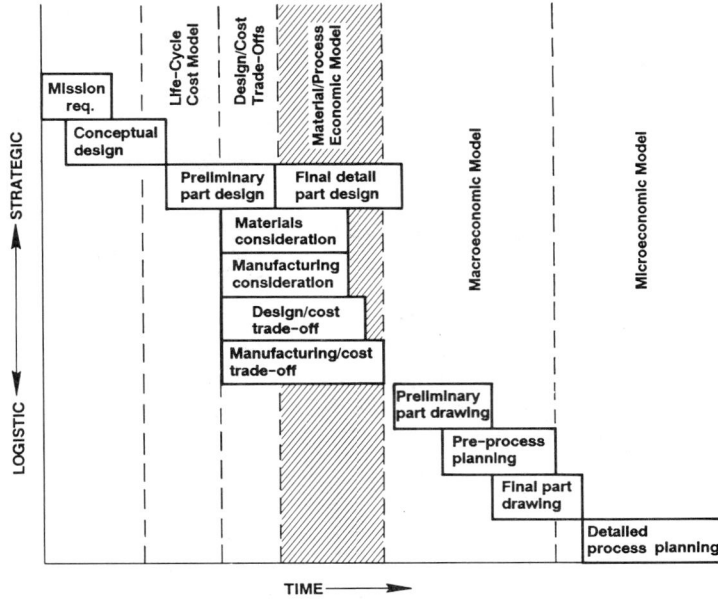

FIGURE 10   Design and manufacturing process development.

Traditionally, most process development projects have concentrated on the establishment of conceptual and technological feasibility. Although this appears to be a logical approach, it is not uncommon to find that a substantial research and development effort results in a process that is not economically feasible and hence cannot be implemented in production. An approach that allows an early evaluation of economic feasibility during the conceptual and technological stages is shown in Figure 11.

The inputs and outputs to each stage are shown on the left and right of the boxes, respectively, and the constraints are shown at the top. The first two boxes on the left depict the development and establishment of the conceptual feasibility that is typical of basic research projects. At these early stages, economic opportunity windows can be established from known or potential applications. The next three stages reflect the establishment of the technological and economic feasibility and the first production implementation of the process. In establish-

ing the technological feasibility, it is necessary to demonstrate that the process is feasible within the technological constraints. Establishing the economic feasibility of the process demands that the economic range of operating conditions be known and used to influence the establishment of the directions for process development.

For evaluation of economic feasibility, the following two criteria must be satisfied:

• The necessary condition: The cost savings per part plus the value of time savings per part must be positive.
• The sufficiency condition: To earn a desired rate of return on investment, the sum of the present values (over the periods covering the life of the process) of cost savings per part plus the value of time savings per part, times the parts per period must be greater than the present value of the required investment.

Note that the necessary and sufficiency conditions are applicable to a straightforward one-to-one substitution of processing

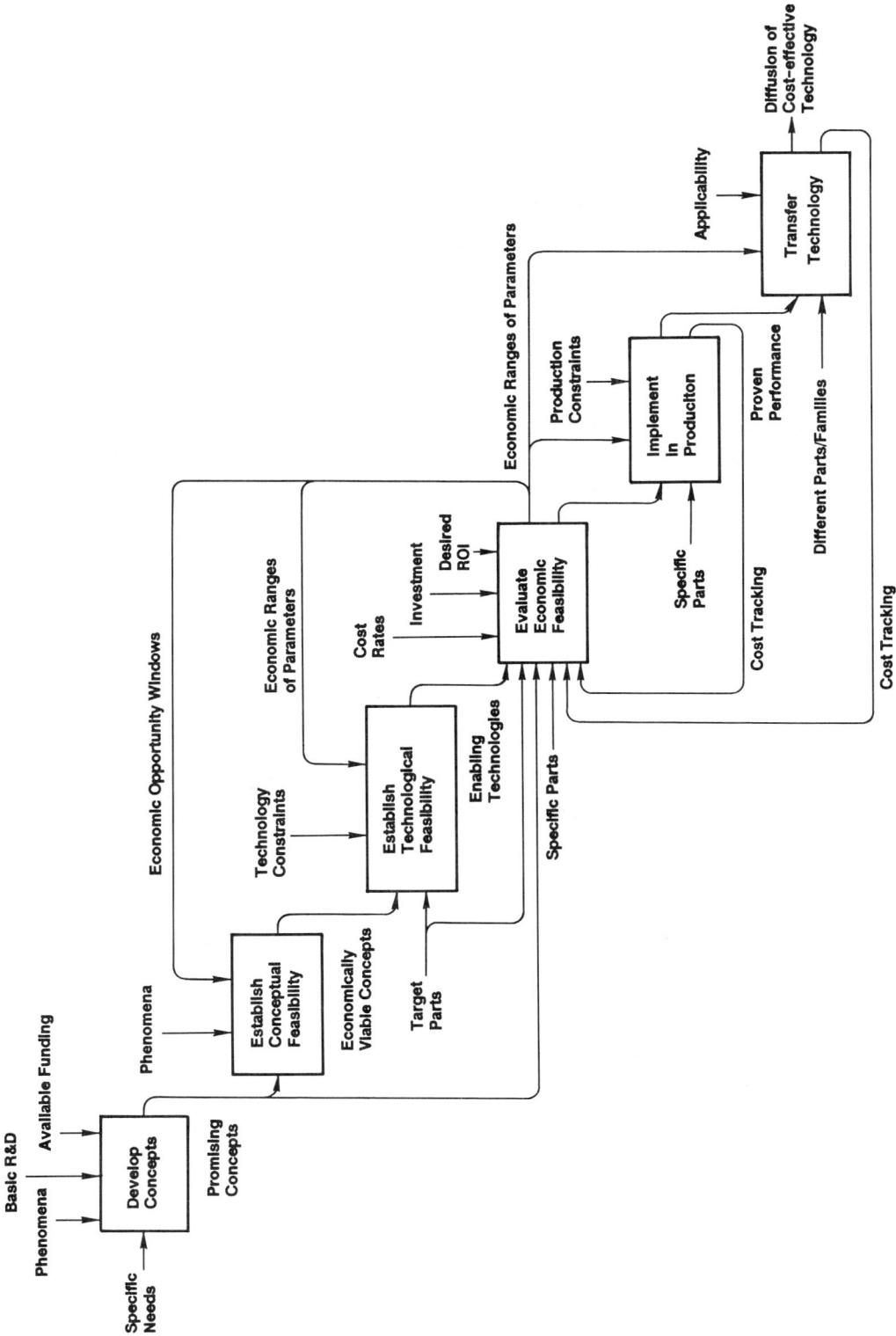

FIGURE 11   Structured analysis diagram showing the stages in the introduction of new technology.

107

units in a sequence as well as to substitutions of the entire sequence of processing units achieving the identical product or service function. Hence, these conditions are useful for evaluating the economic feasibility of replacement as well as new processes and products.

These two criteria, when expressed mathtically, applicable to micro- and macroeconomic models and have been applied to laser-assisted machining, high-speed machining, near-net-shape forging, and process inspection sensor technologies (Tipnis et al., 1981; Tipnis and Watwe, 1983). Furthermore, refinement and application of these criteria should become a powerful aid in guiding emerging technology research and development efforts.

The cost savings of a new process should be evaluated against the best possible current process, since improvement in the current process may negate the expected gains of the new process. An ongoing economic feasibility analysis must answer the following questions:

• What are the economic benefits if the proposed process is found to be technologically feasible?
• What are the targets for process research in view of the opportunities?
• What are true process performance trade-offs?
• What is the sensitivity of these trade-offs and cost factors to economic feasibility of an improved existing process?

The scope of this paper does not permit an in-depth discussion of this important topic (Tipnis et al., 1981; Tipnis and Watwe, 1983). Since process research and development is a time-consuming and expensive activity, a sound methodology to ensure that the development is proceeding toward an economically viable outcome is essential. Further refinement and application of this methodology should be encouraged.

PROCESS PLANNING

Process planning involves the selection of a processing sequence and the operating conditions for each unit process within the sequence that will produce a given lot of parts. The planning can be accomplished only after the initial prove-out of the process. It presumes knowledge in the form of process models or of the processes themselves. Once the preliminary part drawing has been sent to the manufacturing engineer, the activities of preprocess planning, including cost estimation and make-or-buy decisions, are triggered, as shown in Figures 1 and 10. Although make-or-buy and outsourcing decisions are not shown explicitly in Figures 1 and 10, these decisions are typically carried out during the preprocess planning stage. A knowledge of a vendor's capabilities to purchase materials or finished components and to provide proper quality control and testing on the items supplied are critical in creating a coordinated in-house manufacturing and assembly operation with those of the vendor or subcontractor. The overall planning system block diagram shown in Figure 1 has been found to be a valid starting point for the development of a practically implementable computer-assisted process planning system (Tipnis et al., 1979). In this system, "variant" (group technology code-based) and "partly generative" (a graphic generation of operation sequences) has been accomplished.

Although computer-assisted process planning has been an area of active academic research for the past decade, it has not made a sufficient impact on the practice of process planning in industry. This is because most such investigations either have not addressed planning issues across the entire manufacturing sequence or they have made overly severe simplifying assumptions. Thus, the development of most computer-assisted process planning systems in industry has progressed to no more than a glorified word processing and editing system to aid process planners in interactively composing plans.

Since the advent of expert systems, several researchers have attempted to apply them to the problems of process planning. Despite some interesting possibilities, these attempts have revealed that the expert system methodology for planning is not as straightforward as it is for more narrowly defined domains—such as diagnostics—that depend on the expertise of a single individual. Also, the means by which the process knowledge is captured remains largely unexplored (Tipnis, 1987).

Important research areas in process planning include

• Establishment of a methodology for determining processing sequence from specifications of features, dimensions, and tolerance of the part or assembly configurations;

• Application of process models to determine the most economical operating conditions for each processing unit; and

• Evaluation of the impact of operating conditions of each unit process on the production rate and economics of the entire sequence.

Process planning can benefit greatly from a continuing feedback of data from the manufacturing floor. A relational data-base structure appears to be convenient for capturing and reconciling the actual versus planned process operating conditions. Although computer-assisted process planning has been a growing area of research and development, the capability to select an optimal processing sequence and operating parameters remains undeveloped. These topics need to be explored through joint industry and academic research projects.

NEXT GENERATION OF MANUFACTURING SYSTEMS

Traditionally organized manufacturing shops that produce a variety of parts or assemblies in small lot sizes suffer from long lead times, high in-process inventory, excessive reworking and rejects, and high operating costs. To manage this state of complexity, computer-assisted materials requirement planning (MRP) with infinite and finite capacity assumptions has been implemented in a number of U.S. corporations over the past two decades. Also, shop floor data-gathering systems that record the labor hours expended on a given job have been implemented. Although MRP systems assist in achieving economies of scale for material purchasing, they tend to fill the shop with anticipated orders. Experience, however, suggests that unplanned last-minute changes to conform to actual orders have been the rule in most traditional shops.

Modern manufacturing systems have evolved from two separate but complementary directions:

• The development of flexible manufacturing systems with automated material handling has evolved from the transfer line and process flow line concepts combined with the flexibility provided by computer-numerically controlled machine tools.

• The use of just-in-time, total quality control (JIT/TQC) methods in Japan has demonstrated that highly responsive and cost-competitive manufacturing can be done without much capital investment.

These two directions have essentially merged into focused factories that consist of responsive and manageable flexible manufacturing cells and systems. The objective of this section is to identify some of the significant opportunities for research and development related to the role of manufacturing processes within a modern manufacturing system design and operation.

Dependence on Human Skill and Attention

Manufacturing activities, including actual processing and nonprocessing, have been traditionally dependent on human skills and attention. Most tool and die shops continue to depend largely on human expertise for the control of conventional and numerically controlled (NC) equipment. Traditional fixed automation, such as transfer lines, is designed with a built-in sequence of operations that are performed in a predetermined manner. However, the support functions of tool change, readjustment, and maintenance normally depend on manual operations.

Although modern manufacturing systems continue to depend on the skill of the operator for designing, planning, monitoring, and control, the need for constant attention is drastically reduced, thus decoupling the activities of the operator from those of the machines. This enables the process to continue while the operator is attending to other tasks.

In materials processing operations, such as NC machining, once the cut has been initiated, the operator must remain alert to signs of sudden cutter breakage or excessive cutter wear that may throw the dimensions or surface finish out of the prescribed tolerances. An operator experienced in a process can detect the onset of excessive wear by listening to the sounds and feeling the vibrations generated during the cut. Experienced operators also know that it is difficult to stop the machine in time to prevent catastrophic failure of a tool once the failure starts to accelerate. The changes in the chip curl and coloration and in the appearance of the machined surface are sometimes used as indicators of the need to slow down the feed rate to avoid failures. Other manufacturing processes have similar dependence on operator skill to ensure that the process is kept in control.

It is interesting to note that the mundane tasks of chip cleanup, tool loading and unloading and adjustments, and workpiece fixturing are the most difficult to automate to perform flawlessly. These tasks may still require human attention in modern manufacturing systems. When one realizes the extent of human attention necessary for successful operation of discrete parts manufacturing processes, the magnitude of the task of designing and operating unattended manufacturing systems becomes evident.

Manufacturing System Design

Much of the justification for modern automation is based on providing flexibility for rapid setups and changeover to different parts in a predefined family of parts. The concept of flexibility as a focused predefined degree of freedom in a modern manufacturing system can be applied to a wide variety of systems, including small, medium, and large lot sizes and high, medium, and low production rates. Consequently, there is a considerable interest in designing, justifying, and implementing such systems.

The cascade of activities in a plant, as shown in Figure 12, can serve as a starting point for designing a modern manufacturing system. The term *flexible manufacturing system* (FMS) has come to denote a computer-controlled group of numerically controlled machines linked by pallet transport and a load-unload system. Also, tool loading and unloading and tool change activities are under computer control in some flexible manufacturing systems. Since prefixtured palletized workpieces can be loaded or unloaded quickly, the FMS is largely independent of lot size. Whereas the FMS controls the actual production activities, the integrated manufacturing system (IMS) incorporates and synchronizes the preproduction activities of process planning, NC programming, and tool and material acquisition. The IMS, therefore, is a "focused factory" with all the essential preproduction and production functions required to respond to market demands.

The potential benefits of FMS and IMS are shown in Figures 13 and 14, where machine use and throughput time are com-

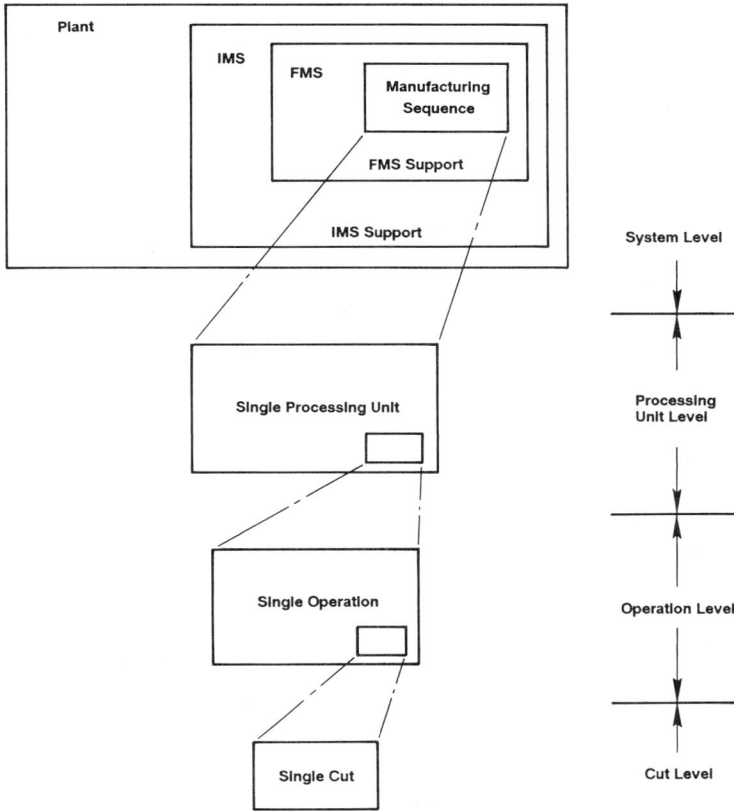

FIGURE 12   Activity levels in flexible and integrated manufacturing systems.

pared with those of the traditional (as-is) manufacturing system. Note that a properly designed FMS provides a significant reduction of nonprocessing times, increases the machine use (that is, the actual cutting or processing time), and reduces the shop throughput time. The IMS significantly reduces preproduction time and hence the total time for completion of an order. FMS and IMS significantly improve the availability of a machine tool to do actual processing, thus increasing the importance of process optimization and control. As noted earlier, processing that proceeds unattended involves guarding against the risks of catastrophic failures and reduced performance.

The first step in designing an FMS is to reexamine processing sequences and operating conditions of the parts in the target family of parts and to evaluate the impact of changes that the processing technology can have on the time and cost of manufacturing. Concurrently, it has proved advantageous to interact with the product designers to evaluate changes and standardization in the part features, dimensions, and tolerances that may lead to significant reductions in manufacturing cost and also reduce the variety of tooling and workpiece fixturing.

The next step is to evaluate all precut and resource supply activities to determine how these can be performed without delaying the actual processing at the workstations.

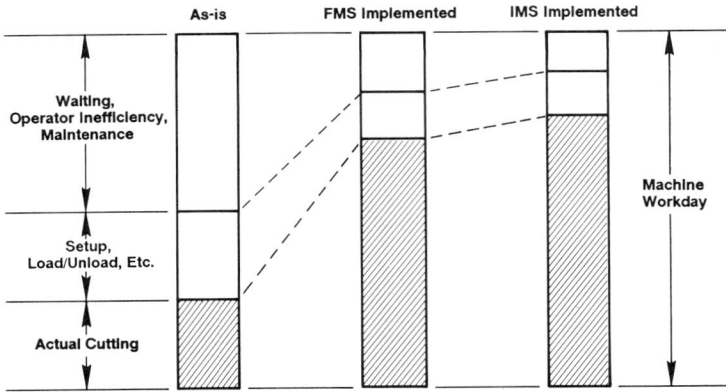

FIGURE 13   Comparison of machine use for a traditional, flexible, and integrated manufacturing system.

Preparation and coordination have been the two prerequisites of a robust design of an FMS.

Another important step is to trade off the cost at each processing unit against the overall system performance. Overall system performance must include measures of production rate, in-process buffer storage, capital investment for all processing units, and the value of work holding pallets and tools. Finally, the system performance can be improved, not by optimizing each processing unit, but by optimizing the processing rate of the bottleneck units and by relaxing the other processing units in the sequence.

Although these manufacturing system design steps are known and practiced by some designers, systematic methodology and design theory are needed. Most design work has been done historically by a few creative individuals in the industry. Most academic research on such systems has followed proven yet narrow tracks of operations research and computer networks. The crucial role of processing in modern manufacturing systems is still not well explored. Still another reason to start the manufacturing system design with the actual process is to avoid the costly risk of obsolescence—that is, investing substantial capital in man-

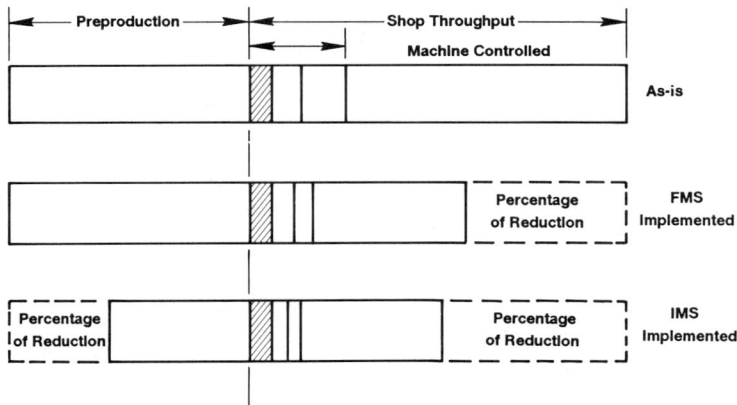

FIGURE 14   Comparison of throughput time for a traditional, flexible, and integrated manufacturing system.

ufacturing processes that may be made obsolete by a new process or a significant improvement in an existing process.

Flexible Manufacturing Systems: Justification and Implementation

The large capital investment and risks involved in FMS have prompted a serious inquiry into the methods of justification and implementation of such systems. A reasonably cohesive methodology that is generic to all FMSs is evolving (Tipnis and Misal, 1985).

The steps involved in this methodology, known as the cost/risk/performance analysis, are shown in Figure 15. The starting point for the analysis is a determination of the strategic advantages of the proposed FMS implementation. Among the key strategic issues are capturing, growing, and maintaining a chosen market share; cost, quality, delivery, and service issues addressed through novel product redesign to gain competitive advantage; and realigning the capital and cost structure through new facilities and outsourcing. It is generally found that an FMS that can be justified from a strategic point of view is more likely to be successful than one that is justified purely on the basis of operational issues, such as increased equipment use, reduced process inventory, and shorter throughput time.

In performing the analysis, it is first necessary to compare the performance levels of alternative FMS designs with those of the current (as-is) manufacturing system and then to determine the potential time and cost savings over the current manufacturing system. At this stage, the microeconomic models for each processing unit and the macroeconomic models for the entire system must be applied. These steps identify potential savings and risk candidates for implementation of emerging technologies. At this stage, lead-time analysis, simulation of material and data flow, and evaluation

of alternative technologies are vital to provide a quantitative basis for comparison of alternative FMS configurations against a more conventional system.

The most crucial step in the analysis is the determination of the degree of risks and the remedies that will overcome the risks that are due to both processing and nonprocessing activities. As stated earlier, the three types of risks that need to be considered for each operation and activity are catastrophic failure, reduced performance, and obsolescence. There is a need to develop a systematic risk analysis methodology similar to a fault-tree analysis but able to contain various degrees of failure. Lacking such a methodology, the risks are evaluated by subjective probabilities obtained through the consensus of experts.

Another important step is a comparison of the cost allocations to the processing units and support activities in the FMS. Traditional accounting and financial methods are not sufficiently refined to allow such allocation. Also, the distribution of overhead is traditionally based on labor hours. For automated systems, an allocation based on machine hours would appear to be more appropriate. Because most FMSs can significantly save processing and throughput time, the value of the time saved from actual processing as well as the value of reduced scrap and reworking should be considered. Most traditional accounting methods do not provide for such time-value-based savings.

Clearly, a comprehensive methodology for cost allocation is needed as a part of the FMS economic model. Initial efforts in this direction show that the foundation for a sound economic model requires advances in the economics of technological change, a subject not well explored to date.

CHALLENGES AND OPPORTUNITIES

The central role that processing plays in the manufacturing system has been empha-

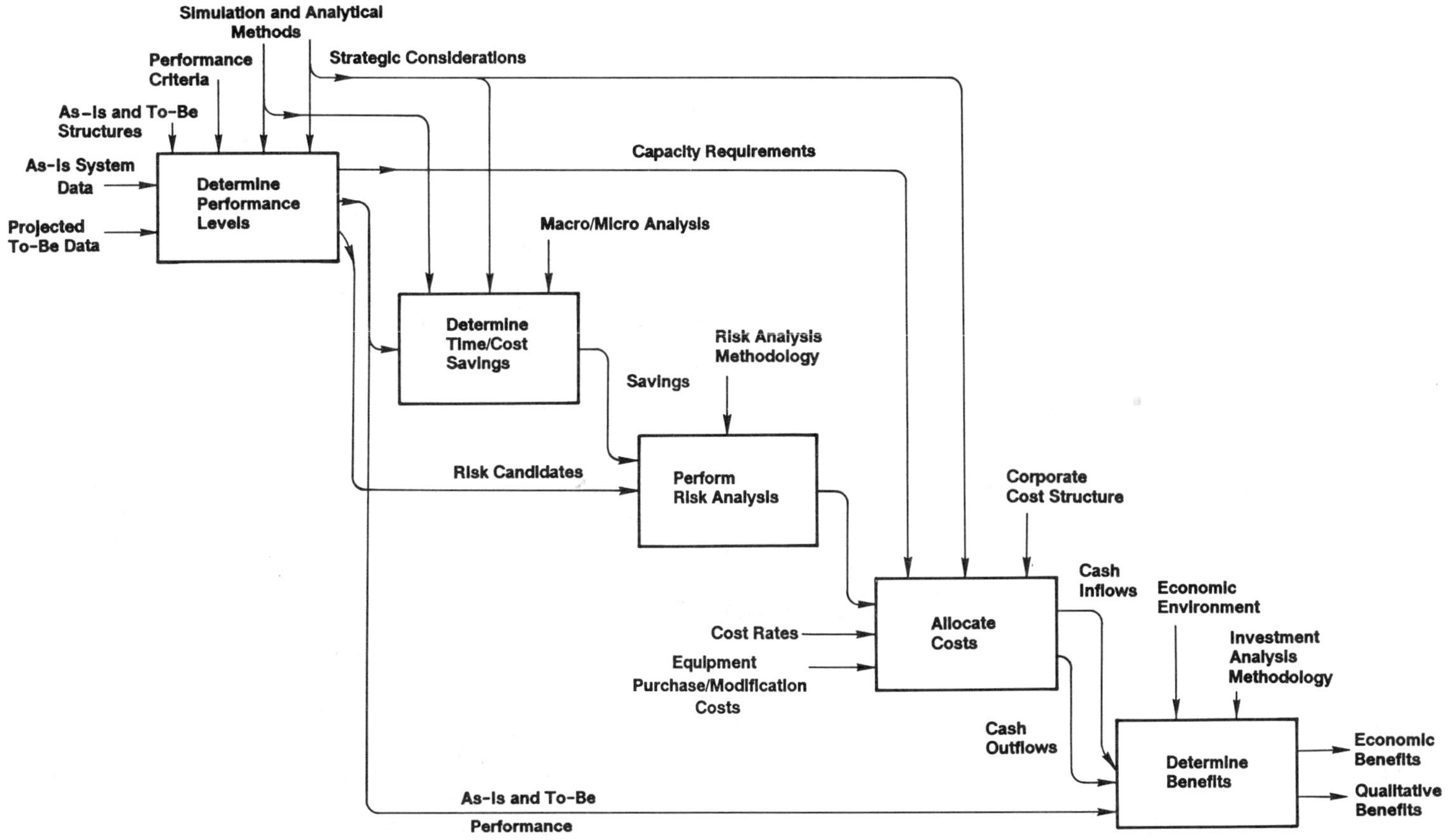

FIGURE 15 Cost/risk/performance methodology for FMS/IMS justification.

sized throughout this paper. Although the role of other nonprocessing activities (for example, data flow or material handling) are important, one must not forget that the purpose of the manufacturing system is to create a product. Robotics, vision and other sensors, and expert systems have been emphasized in recent academic research projects. Although these topics do represent ripe areas for exploration, their limited focus rarely allows advances in design and operation of a total manufacturing system.

Manufacturing systems have evolved from the traditional functional organization derived from the so-called scientific management of F. W. Taylor and others during the early 1900s to today's highly focused factories. The functional organization depended on the efficiencies derived from specialization of labor and machines and standardization of manual tasks. However, the flow of work through a series of functional workstations made the flow of parts complex and lengthy. The functional emphasis also removed the responsibility for overall part quality and throughput from any single workstation operator. Each operator became responsible for only the work done at that workstation. Thus, the overall system—the combination of all elements of the transformation process from the initial design to the final product—was sometimes not visible. Throughout this paper, concerns have been noted with the current understanding of the physical processes, with the models of the process, with the economics of the system, and with the overall system models.

The central problem is that the physical understanding of most discrete parts manufacturing processes is not sufficiently refined for practical application of phenomenological process models. Not only do these areas deserve more attention, but also different ways of attacking the problems are needed. For example, materials and materials processing research and development are frequently located in different aca-demic departments in the university, making interdisciplinary research difficult. In industry, most of the new process development focuses on near-term problems, and limited attention is being given to investigations of the underlying phenomena. Increased cooperation between industry and university should lead to the identification of problems of highest priority and should improve the overall effectiveness of materials and materials-processing research.

Basic research on process models and methodologies must be pursued on an ongoing basis. As was noted earlier, it is of great importance that current limitations in these theories be reduced so that they can be more broadly applied—for example, to those areas in which the microstructure of the materials is treated in detail and the typical processing conditions of extreme stress, strain, strain rate, and temperature conditions can be accommodated.

The state of the art of empirical process models should be advanced to develop a sound methodology for constructing models of practical use. Currently available models have proved to be implementable for process control and process optimization. However, a strengthened mathematical basis is needed for constructing algorithms of objective functions and the constraints for a variety of materials-processing operations. Furthermore, it should be recognized that the best quality control is to control the process so as not to produce parts and assembly configurations outside the prescribed quality limits. Research is needed not only to develop on-line and look-ahead sensor-based process control but also to establish process control strategies and algorithms based on tolerance requirements for parts and assemblies.

Influencing the design to achieve periodicity is an important unsolved problem in those instances in which the essential communication media are drawings that cannot, as yet, be guaranteed to be free of ambiguity, incompleteness, or inconsistency of features, dimensions, and tolerances. Process capabilities also need to be captured in representations suitable for establishing process alternatives and design and manufacturing cost trade-offs.

The related subject of computer-assisted process and operation planning has attracted much attention for the past two decades. Much work remains to be done to make it possible to generate process and operation plans from basic principles. Capturing processing expertise that cannot be formalized into mathematical models is still a major challenge.

The economics of unit manufacturing processes and sequences and manufacturing systems is finally receiving the attention it deserves. A firm foundation is needed to establish economics as the guiding force for design, development, planning, optimization, and control of processes and manufacturing systems.

Economic criteria should also be able to identify those new processing technologies that offer the greatest opportunities for addressing the challenges of the 1990s. To make wise use of the limited research and development resources and talents for the selected few opportunities is the key to being competitive in manufacturing in the 1990s.

REFERENCES

Bjorke, O. 1978. Computer-Aided Tolerancing. Trondheim, Norway: Tapir.

Clark, J. P., and M. C. Flemings. 1986. Advanced materials and the economy. Scientific American 255(4):51–57.

Compton, W. D., and N. A. Gjostein. 1986. Materials for ground transportation. Scientific American 254(10):93–100.

Ford, H. 1966. Unsolved problems associated with the forming of metals: Metals transformations. Pp. 193–210 in Proceedings of the 2d Buhl Conference, W. W. Mullins and M. C. Shaw, eds. New York: Gordon & Breach.

Hillyard, R. C., and I. C. Braid. 1977. The Analysis of Dimensions and Tolerances in Computer-Aided Mechanical Design. Computer Laboratory, CAD Document No. 93. Cambridge, England: Cambridge University.

Kapoor, S. G., and S. M. Wu. 1980. DDS with applications to manufacturing processes. Pp. 403–414 in Advanced Manufacturing Technology, P. Blake, ed. International Federation of Information Processing (IFIP). Amsterdam: North-Holland.

Light, R. A. 1979. Symbolic Dimensioning in Computer-Aided Design. M.S. thesis. Massachusetts Institute of Technology.

Novak, E. 1980. Adaptive Control. Ph.D. dissertation. Royal Institute of Technology, Stockholm, Sweden.

Opitz, H. 1966. Unsolved problems associated with metal removal operations: Metal transformations. Pp. 261–305 in Proceedings of the 2d Buhl Conference, W. W. Mullins and M. C. Shaw, eds. New York: Gordon & Breach.

Ravignani, G. L., V. A. Tipnis, and M. Y. Friedman. 1977. Cutting rate-tool life functions (R-T-F): General theory and applications. CIRP Annals 25(1):295–301.

Requicha, A. A. G. 1977. Dimensioning and Tolerancing. Report T-M19, Production Automation Project. University of Rochester, New York.

Shaw, M. C. 1966. Historical aspects concerning removal operations on metals: Metal transformations. Pp. 211–260 in Proceedings of the 2d Buhl Conference, W. W. Mullins and M. C. Shaw, eds. New York: Gordon & Breach.

Shoemaker, W. W. 1980. Life cycle cost as a tool in the detail design of advanced propulsion system. AIAA/SME/ASME 16th Joint Propulsion Conference.

Smith, B. 1987. PDES First Testing Draft. National Bureau of Standards, A101 Bldg. 223, Gaithersburg, Maryland.

Taylor, F. W. 1907. On the art of cutting metals. Transactions of the American Society of Mechanical Engineers 29.

Tipnis, V. A. 1977a. Mathematical models and algorithms for adaptive control of NC end milling operations. Pp. iv-8–iv-18 in Proceedings of the International Conference of Production Engineering, New Delhi, India. Calcutta: Institution of Engineers of India.

Tipnis, V. A. 1977b. A strategy for the development of improved machining of steels. Pp. 1–4 in Proceedings of International Symposium on Influence of Metallurgy on Machinability of Steel. Tokyo: Iron and Steel Institute of Japan.

Tipnis, V. A. 1987. Computer-aided process planning: A critique of research and implementations. Pp. 295–300 in CIRP Manufacturing Systems Seminar, Pennsylvania State University.

Tipnis, V. A., and A. C. Misal. 1985. Economics of flexible manufacturing systems. SME paper MS 85-

154. Dearborn, Mich: Society of Manufacturing Engineers.

Tipnis, V. A., and U. Watwe. 1983. Economic models for processing alternatives, I.—Relationship between process, economic and life cycle cost models for near net shape parts. Pp. 131–174 in Experimental Verification of Process Models. Metals Park, Ohio: American Society for Metals.

Tipnis, V. A., H. L. Gagel, and S. A. Vogel. 1978. Economic models for process planning. Pp. 379–387 in Proceedings of the Sixth North American Metalworking Research Conference, Gainesville, Fla.

Tipnis, V. A., S. A. Vogel, and C. E. Lamb. 1979. Computer-aided process planning system for aircraft engine rotating parts. Pp. 151–169 in Society of Manufacturing Engineers Computer and Automated Systems Association Technical Paper MS 79-155. Presented at the Prolomat Conference, Ann Arbor, Michigan.

Tipnis, V. A., G. L. Ravignani, and S. J. Mantel, Jr. 1981. Economic feasibility of laser assisted machining. Pp. 547–552 in Proceedings of NAMRAC IX Conference.

Tipnis, V. A., S. J. Mantel, G. L. Ravignani, and U. Watwe. 1984. Economic Modeling. Advanced Machining Research Program, Vol. 5, Report No. TR-84-4059. Air Force Wright Aeronautical Laboratory, Wright-Patterson Air Force Base, Ohio.

Voelcker, H. B. 1988. Modeling in the design process. In Design and Analysis of Integrated Manufacturing Systems, W. Dale Compton, ed. Washington, D.C.: National Academy Press.

Whitney, D. E., J. L. Nevins, T. L. DeFazio, R. E. Gustavson, R. W. Metzinger, J. M. Rourke, and D. S. Seltzer. 1988. The strategic approach to product design. In Design and Analysis of Integrated Manufacturing Systems, W. Dale Compton, ed. Washington, D.C.: National Academy Press.

Wu, S. M., and D. S. Ermer. 1966. Maximum profit as the criterion in the determination of the optimum cutting conditions. Transactions of the American Society of Mechanical Engineers, Series B, 88(4):435-442.

A NEW PERSPECTIVE ON MANUFACTURING SYSTEMS ANALYSIS

Rajan Suri

ABSTRACT   Most U.S. corporations are devoting tremendous efforts to revitalizing their manufacturing base. As a result of the enormous effort in planning, designing, evaluating, and operating this vast array of plants, manufacturing systems modeling has become a vital activity. Coincident with this major change in facilities, the manufacturing field has itself been undergoing radical changes. On the one hand are the high-tech developments enabling automation and integration, and on the other are philosophical developments such as just-in-time. Much of the "traditional" analysis activity has not been directly relevant to these changes, and the major recent thrusts in this area have come from outside the traditional community. The role of analysis in modern manufacturing is discussed, research directions are proposed, and steps to be taken by the individual and the research community as a whole are suggested.

INTRODUCTION

A tremendous effort is being devoted to restructuring the manufacturing base of most industrial corporations. In the wake of the resultant planning, design, evaluation, and operation of vast numbers of plants, the analysis and modeling of manufacturing systems has become a vital activity. But as this interest in manufacturing analysis has grown, those of us who have concentrated on "traditional" analysis approaches must ask whether we are in the mainstream.

This paper looks at some recent developments in this field and contrasts them with traditional approaches. It presents the hypothesis that the recent principal contributions in manufacturing analysis have come from outside the traditional communities. It proposes a framework for viewing the

issues in manufacturing analysis and suggests some research opportunities. It proposes a novel concept, design for analysis, for treating this area. It concludes with research suggestions for individuals who wish to make useful contributions in this area.

Throughout this paper, the term "analysis" is used generically to include both analysis and modeling, and the term "traditional" analysis methods means analysis and modeling approaches that were traditionally found in the publications of societies such as APICS, IEEE, IIE, ORSA, SME, and TIMS up through the early 1980s. Examples of such publications are *IIE Transactions*, *Management Science*, and *Operations Research*. In addition, the adjective "analytic" is used exclusively to describe techniques that rely on solving equations, while the term "analysis" is used to describe broadly any technique for systematically studying a problem. Thus, in this paper, simulation would be an analysis technique, not an analytic technique.

"TRADITIONAL" MOTIVATION FOR MANUFACTURING ANALYSIS

We begin by reviewing some of the traditional arguments for conducting manufacturing systems analysis. Manufacturing systems analysis and performance evaluation is not a new field, as any industrial engineer will testify. Throughout the lifetime of a typical system, the organization responsible for it goes through many phases of decision making, from an analysis of initial feasibility through design, operation, and finally obsolescence. Typical decisions that must be made in a manufacturing environment and typical performance measures used to evaluate these decisions are shown in Table 1.

A modern manufacturing system, however, can be quite different from the more traditional system. A modern manufacturing unit is most often a complex system, consisting of many interconnected components of hardware and software. Decision making, therefore, can become difficult because of the greater complexity of the modern system compared to, say, a conventional job shop. This complexity is due to several factors:

• Highly interconnected components leading to a very large set of decisions that must be made simultaneously. In the modern system, where both material and information move rapidly through the plant, a small change at one end of the plant can have, in a few minutes, a significant impact on an area at the other end of the plant.

• Limited resources due to efficiency requirements. The sharing of resources as a means of reducing costs increases the complexity of managing that sharing and of predicting simultaneous demands for shared resources.

• Little "slack" in the system. A commonly quoted statistic for batch manufacturing typical of job shops is that machine tools spend 5 percent of their time performing value-added tasks—e.g., cutting metal—so these conventional systems have a lot of "slack." In an automated system, such as a flexible manufacturing system (FMS), this number can be 10 times higher or more, leaving little slack (U.S. Congress, Office of Technology Assessment, 1984).

• Fewer "humans in the loop." Human operators can use common sense to correct or modify situations caused by unexpected changes in conditions. In a highly auto-

TABLE 1 Typical Performance-Related Decisions During the Design and Operation of a Manufacturing System

Typical Decisions	Typical Measures of Performance
Number and types of machines	Period of payback
Number of load/unload stations	Return on investment
Part-types	Net present value
Alternative routings	Facility utilization
Tool allocation	Production rates
Number and types of fixtures	Work in progress
Number of transporters and pallets	Part and material flow times
	System flexibility
System layout	Queues at each resource
Buffer sizes	Operating policies

mated environment, these unexpected perturbations must have been anticipated and provided for or the control system will not be able to properly react. Disaster can result from inadequate software controls.

Because of all of these factors, even experienced shop floor supervisors and managers have difficulty in perceiving all the consequences of any given action in the modern manufacturing context. Also, because of such factors, we found that, in the words of an experienced FMS user, "People thought they were getting into FMS; instead they got into FM-Mess!" (J. Schnur, personal communication, 1985).

These factors have also created a genuine dilemma for the designer of manufacturing systems. On the one hand we have complex systems involving high risk and high capital, yet on the other we have the requirement to design and operate these systems efficiently to achieve the corporate objectives of competing effectively. It is precisely to help resolve this dilemma that the traditional analysis community entered the manufacturing arena. A typical reaction, including that of the author, was that the availability of tremendous computing power made it possible to build large models to analyze the system and aid in the decision making. Researchers in this community have been publishing increasingly complex analyses that had "application" to manufacturing. The journals are currently filled with articles that address complex modeling issues for such things as FMS, robotic cells, and automated storage and retrieval systems (AS/RS). But what should be the priority afforded all of this research effort that is devoted to modeling and analysis? It will be informative to see what the competition has been doing in these areas.

THE "JAPANESE WAY"—A DIFFERENT APPROACH

The Japanese approach to manufacturing has fundamentally affected the way in-

dustry views this field (Hall, 1983; Schonberger, 1982). There has been a corresponding effect on manufacturing analysis. It is helpful to contrast the approaches of most American manufacturers and the Japanese manufacturers to three common features of many systems.

Complexity

The American approach to dealing with large complex manufacturing operations has been to adopt material requirements planning (MRP) and manufacturing resource planning (MRP-II). A little introspection leads us to realize that these are just large models, with various assumptions of lead times, lot sizes, demand, etc. The Japanese approach is to simplify the operation rather than to attempt to deal with the complexity. Examples of this are the "flow line," which simplifies the product flow, and the *kanban* system of job tickets, which simplifies shop floor scheduling and control.

Uncertainty

The American approach to countering uncertainty in the manufacturing environment has been to use buffer stocks and lead times with imbedded safety margins. These safety stocks or safety times have been derived by means of modeling techniques (e.g., economic order quantity, or EOQ). The Japanese way of tackling uncertainty is *to get rid of it:* the just-in-time (JIT) philosophy espoused in numerous recent articles has, as one of its goals, the systematic reduction of uncertainty in all aspects of the operation.

Constraints

Manufacturing managers are constantly dealing with constraints—constraints on capacity, on tooling, on precedence of operations, etc. The American way of dealing with these constraints has been to develop more and more sophisticated models that attempt to optimize schedules within numerous complex constraints. Once again, the Japanese have come forth with a radical approach to constraints—*break them!* Although this may sound frivolous, there are many instances of success. Proponents of the Japanese approach are eager to cite the case of setups at large stamping presses in the automobile industry. Typically, American manufacturers would take around 10 hours to change the setup for a different component. To minimize the number of tooling changes, large lot sizes were run. This constrained the schedules of preceding and following operations and also involved optimization to find the best inventory levels for the system as a whole. Finding the best lot sizes was a complex problem that had to be solved each month, to say nothing of the system-wide effects that resulted from any changes in schedule.

The Japanese tackled the problem differently. Studying the setup procedure itself, they found engineering and equipment solutions to shorten the procedure, eventually achieving tooling changeovers that consumed only a few minutes. Since the amount of time required to change a die was now negligible, it no longer mattered what sequence or lot sizes were needed and the setup and lot size constraint simply disappeared.

The point of this example is to highlight the differences in the two approaches, as summarized in Table 2. Essentially, the American approach has involved taking various problems as given and developing models to deal with those givens. The Japanese approach has been to try to eliminate the problems. In the light of the success of the Japanese techniques, and the rate at which American industry is attempting to adopt them, does this mean we can do away with traditional analysis of manufacturing systems? We will return to this question later.

OTHER "NONTRADITIONAL ANALYSIS" SUCCESS STORIES

Recently, there have been three other highly visible thrusts in the area of manufacturing systems analysis.

Artificial Intelligence and Expert Systems

Artificial intelligence (AI)-based approaches, particularly expert systems (Fox and Smith, 1984), have been receiving much enthusiastic attention from industry. Indeed, one gets the distinct impression that an AI-based project would be more likely to receive funding or contracts than one based on traditional approaches.

Optimized Production Technology

Optimized production technology (OPT) is a system that has quickly gained a lot of visibility in the area of manufacturing systems analysis. OPT is both a software system and a philosophy (Lundrigan, 1986; Meleton, 1986). The software system has

TABLE 2 Two Cultures of Manufacturing Systems Analysis

Problem	Approach	
	The American Way	The Japanese Way
Complexity	Models—MRP and MRP-II	Simplify (e.g., flow lines)
Uncertainty	Models to derive lead times and buffer stocks	Get rid of it (e.g., JIT)
Constraints	Models to optimize	Break them (e.g., setups)

the ability to schedule very large factories. To use the software one has to adopt the OPT philosophy, much of which is reasonable and simple. Although one might debate some of the ideas in OPT, one fact seems clear—in a short period of time, this system has shown that there exists a tremendous market opportunity for effective analysis and scheduling software. A typical OPT installation may require investments exceeding a million dollars.

Animation

Computer simulation with graphic animation is one of the increasingly popular tools used in industry for analyzing manufacturing systems. It is widely accepted that the animation component helps to convince both manufacturing engineers and senior management of the benefits of simulation. The popularity of this approach can once again be judged by the growing number of commercial software systems available for the task (Haider and Banks, 1986). Although animation is having the positive effect of getting simulation to be used, it is unfortunately also having a negative affect. It is becoming a substitute for simple analysis. In fact, one sees conference presentations where traditional analysis is replaced by poorly analyzed, but very pretty, graphic animation studies. One hears of major projects being funded on the basis of such studies, where the basic assumptions about the data and approach were flawed but the animation sold the project.

Summing up the New Developments

Four of the most recent and visible developments in manufacturing analysis are the following:

- The Japanese approaches
- AI/expert systems
- Optimized production technology
- Graphic animation

These are prominent recent approaches in the sense that they appear to be receiving more attention, publicity, and money than any other analysis techniques. The surprising, indeed shocking, fact is that none of these approaches reflects developments from the traditional analytic community. One can raise many questions concerning this observation. Is this an indication of the inability of the traditional approaches to be implemented effectively? Is it an indication of an inability to sell concepts effectively? Or, should we, the traditional experts, give up working in this application area? Although specific answers to each of these questions will not be proposed, the balance of this paper attempts to identify new directions for the traditional analyst that offer the possibility of reversing this trend. At least, it is hoped that raising some of these questions will lead to further introspection about where this community, the traditional analysis experts, is going, where it needs to go, and how it should direct its efforts in getting there.

ROLE OF ANALYSIS IN TODAY'S CONTEXT

Recent Developments

There *is* a role for traditional analysis techniques in the context of new developments. As an example, let us consider the Japanese approach of just-in-time.

- A point often lost in the publicity surrounding these methods is that the Japanese approach is an ideal. Although one must constantly strive to achieve it, we will never live in a perfect world such as one with no uncertainty. Real-world companies face constant change—in demand, in technology, in people—so one must continue to pursue improvements.
- No corporation can implement the Japanese approach overnight—they need to "get there from here."
- It is recognized that the Japanese ap-

proach may not be suitable for certain products and certain distribution of facilities and suppliers. Thus, some facilities will continue to be run in the more traditional "American style."

Now reconsider these points, but in a logical order, starting from the final point above. Is the Japanese approach suitable for a given set of products? Traditional analysis methods can help answer this question. How do we get there from here, given that we would like to maintain a reasonable delivery schedule with our customers even as we implement JIT? Again, traditional analysis can help us anticipate some of the problems before they occur and can aid in maintaining deliveries. To use the familiar Japanese analogy, just as the JIT approach lowers the "water level" (the inventory) to expose the rocks (the problems), so analysis can act as a "sonar" to alert one to distant rocks. As the real world changes around us, various types of traditional models can serve as decision aids to understand the impact of alternative choices and varying changes in the environment, even while JIT is being used.

In a similar way, traditional analysis methods can be used to complement or enhance other recent analysis approaches such as expert systems. The challenge is to show that the traditional analysis community can successfully learn what is needed and produce effective implementations and success stories similar to the four described in previous sections.

Keys to Effective Analysis

A key to understanding the current trends in the general area of manufacturing is to recognize that manufacturing is now regarded as a major weapon in a company's strategic arsenal (Hayes and Wheelwright, 1984; Skinner, 1985). This must be constantly recognized during the analysis process. The manufacturing system analysis process, as shown in Figure 1, must be driven by management's strategic objectives. Thus, one of the keys to effective analysis is a recognition of the drivers of this process. A second key to effective manufacturing system analysis can be appropriately stated by paraphrasing the just-in-time statement: *Use the right model at the right time to answer the right question.* These two key points of effective manufacturing systems analysis have often been forgotten in the voluminous "traditional" publications of industrial engineering, management science, and operations research.

Even if one agrees, in principle, with the foregoing statements, the question arises as to how to identify the right questions and models. At a given point in time, what should one model? What decisions and parameters should be included? What performance criteria should be evaluated?

We propose that a systematic structure for answering these questions is obtained by looking at the "manufacturing system life cycle."

LIFE CYCLE PHASES, ANALYSIS, AND RELATED RESEARCH ISSUES

Manufacturing System Life Cycle

The life cycle of an item is defined as the period from the initiation of the concept to its obsolescence. Both products and systems have life cycles. While the product life cycle is a widely studied concept, the life cycle of modern manufacturing systems is not. With the life cycle of modern products getting

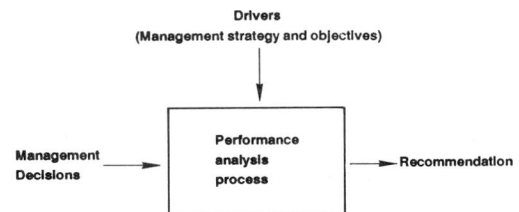

FIGURE 1   Representation of the manufacturing system analysis process.

shorter—e.g., two to three years for some electronics products—there is a major motivation to create manufacturing systems with a life cycle that spans several product life cycles.

Although many factors determine the life cycle of a manufacturing system, the emphasis in the following discussion is on those aspects that are most amenable to analysis. Even though strategic analysis is a critical first step in the determination of business goals, markets, and products, this must be accomplished before any substantial conceptualization of the manufacturing system. We begin the following discussion with the feasibility analysis stage. Each of the following sections discusses some of the critical stages in the manufacturing system life cycle, identifies appropriate analysis needs, and suggests selected research issues.

Feasibility Analysis (Planning) Phase

The objectives of the feasibility analysis phase are to establish the economic attractiveness of various system alternatives that are candidates for the production of one or more of the stated products. It will be necessary to obtain initial estimates of various system measures, as described later, for each of these alternatives. Typical issues that must be addressed during this phase are as follows:

• An understanding of the organizational objectives that this system is intended to meet.
• A determination of the geographic location of the manufacturing facility.
• A determination of the products, subassemblies, and components that are to be made or bought.
• An understanding of the impact of alternative product designs on the manufacturing process and system configuration.
• An understanding of the implications of alternative process approaches, such as "FMS" versus "flow line."

• The identification of candidate equipment choices for each alternative.
• A decision concerning the degree of "flexibility" that the system should have.
• An understanding of the constraints that will be imposed on the issues by different levels of capital investment.

Several measures of performance must be applied to each of the system alternatives that are developed in this analysis. Typical measures of performance for the capital that will be invested are the return on the investment (ROI), the net present value (NPV), and the payback period. A measure of the strategic return will involve an assessment of the impact on quality, price, and responsiveness to the marketplace. The ability to respond to product changes, to demand surges in the marketplace, and to operational uncertainties must be estimated.

Typical analysis techniques that are appropriate for this phase are forecasting, decision analysis, location analysis, economic analysis, mathematical programming (e.g., linear or integer programming and nonlinear programming), facility layout, and group technology. Detailed references on these and a few other appropriate techniques are available (Suri, 1985). Although it appears that traditional analysis methods are applicable for this stage of analysis, the literature and methodology are deficient in essential areas that are relevant to current manufacturing practice. Large, complex mathematical programs to study optimal location of a facility exist, but there is little available that can be used to characterize the impact of strategic alternatives on quality and responsiveness.

The following list of research issues is not exhaustive, but it is indicative of the problems that remain. An additional recent perspective on this topic can be found in Gershwin et al. (1986). It will be obvious that the problems identified can be successfully addressed only with new tools. It is the development of these tools that will form the basis for the research tasks.

• Better decision models for evaluation of investments in rapidly evolving technology. Companies in the semiconductor industry regularly confront this problem. They must make decisions concerning new factories that may not be in full production for four years, which means making decisions well before the time that the products or processes have been fully specified.

• Development of integrated approaches to product, process, facility, and equipment decisions. As elaborated by Whitney et al. (1988, in this volume), the benefits of integrating these decisions can be substantial. The research challenge here is to do this more efficiently than is possible with large and complex computer models.

• Improved quantitative understanding of the benefits of properties such as responsiveness, product quality, and system flexibility and how these properties affect market share and profitability.

• Improved frameworks for quantifying the benefits of investments in flexible facilities and for assessing the appropriateness of various types of flexibility.

• Provide answers to questions such as "Should this plant implement JIT?" and, if so, "Where should inventory be reduced, by how much, and when?" A gradual introduction of JIT is sometimes desirable. An understanding of alternative programs of implementation is needed (Suri and DeTreville, 1986).

Aggregate Analysis Phase

The aggregate analysis phase arises after the feasibility analysis has shown that the project is worth pursuing and management has given the go-ahead for further design and analysis. The objectives of this phase are to further evaluate the systems selected in the previous phase, to design a rough configuration for each candidate system, to reduce the number of alternative systems (typically to one or two), and to verify that the assumptions made in the feasibility analysis phase are still appropriate in the light of this additional analysis. Typical issues to be addressed during this phase are a determination of the processing capacities that will be needed at each stage of the operation; the effect of various parameters such as equipment reliability, process yield, and product volume and mix; the impact of alternative lot size choices on the processing parameters; the approximate requirements for material handling and storage; and the requirements for computer hardware and software.

Various measures of performance are appropriate for this stage of analysis. These include refinement in the values of all the measures used in the earlier feasibility stage as well as measures of the product cycle times, the work in process (WIP) inventory, the size of queues, the level of equipment use that can be expected (including the expected downtimes), the performance of the computer hardware and software, and the expected production rates that can be achieved with the assumed production configuration. At this stage the analysis should be done at a "high" level, with many details still being approximated or aggregated. For example, it would be appropriate to consider material handling and storage equipment at an aggregate level and to delay any detailed scheduling and sequencing considerations. Various aspects of aggregate analysis have been discussed elsewhere (Suri, 1985; Suri and Diehl, 1987). The effective use of the aggregate analysis approach for a modern factory was described by Haider et al. (1986).

Typical analysis techniques that will be found useful in the aggregate analysis phase are strategic and economic analysis, such as decision-tree analysis and discounted cash flows, the use of queueing analysis and queueing network models, simulation models that can treat aggregated systems (Suri and Diehl, 1987), models for the reliability of machines and processes, models that project the yield from individual processes, and models that provide a measure for computer systems performance.

Selected research issues that relate to aggregate analysis are these:

• Development of improved analytical models—e.g., queueing models and reliability models—that can treat situations at an appropriate level of aggregation. The aim of such models should be to perform easy and quick high-level analysis of system alternatives. This approach implies that the models should be simple even at the expense of precision. This contrasts strongly with the direction of much of the research effort devoted to developing intricate refinements that make queueing models more accurate—e.g., reducing the error from 15 percent to 2 percent. Given the role suggested here for such models, an accuracy of 15 percent is acceptable. The simplicity and improved speed of solution that can result from the less detailed models can be significant advantages. Thus, in contrast to continuing the search for further refinements, we pose the following key problems in this area: (a) the ability to model "blocking," such as occurs when limited buffers are available or when there is an attempt to simultaneously share a resource; (b) the development of a better understanding of the systems and conditions under which such models work well and the conditions under which they should not be used; (c) the development of improved analytical models for transient analysis—e.g., start-up after a failure; and (d) the development of more validation studies based on the use of these models for real manufacturing systems.

• Models that will provide an improved understanding of the aggregate dynamics of *kanban* systems.

• An improved understanding of the overall impact on quality and yield through integrated models of equipment perfor-

mance, system dynamics, and economics.

• Improved algorithms for optimization of simulations.

Detailed Analysis Phase

Earlier analysis phases will have narrowed the candidate systems to one or two choices. The objectives of this phase are to establish, in as much detail as possible, how the candidate system(s) will function, including how the system will operate in concert with the rest of the factory and the rest of the organization. It will be necessary to verify that assumptions made during the earlier analysis phases remain valid in the light of the increasing level of detail that is now available. The most desirable system must now be selected, along with decisions on all of the relevant parameters. Typical issues to be addressed in this phase are the detailed configuration for all equipment, including tooling and fixturing requirements; the explicit details and characteristics of the material handling and storage system; an understanding of the "links" to the rest of the organization (e.g., suppliers, MRP system, shipping, and other logistics); the mechanism for translation of corporate requirements into individual tasks for equipment in this system; the determination of effective planning and scheduling policies; the determination of the system response to short-term problems such as failures, blockages, and late arrival of material; the determination of the management structure and labor structure; and the definition of functional requirements for all computer hardware and software elements.

Performance measures that are appropriate for this phase include those used in previous phases plus the utilization levels and response times for the material handling and storage subsystem, the utilization levels of the tooling and fixturing, the frequency of blocking and resource contentions, the response times for the computer hardware and software subsystems, and the reliability and responsiveness of the overall system to problems. Although the typical analysis techniques used in this phase include all those used in previous phases, one finds in practice that the technique most used here is discrete-event simulation. Additional techniques that are found to be useful, however, are production planning, lot sizing, scheduling and sequencing, real-time control, and structured system analysis and software design.

The extensive literature on mathematical programming methods for lot sizing and scheduling is simply not appropriate in the current context of simplified cellular manufacturing and JIT operation. Although simulation is widely used by industry, few practical tools exist for the statistical analysis of simulation output, and almost none exist for the optimization of simulation results. Hence, in the framework suggested in this paper, selected research issues relevant to the detail analysis phase are these:

• Development of a theory of discrete event dynamic systems (DEDS). A much better qualitative and quantitative understanding of the detailed operation of manufacturing systems should result from a good theoretical foundation for DEDS. There has been some progress in this area, such as developing a linear system representation (Cohen et al., 1985), formalizing notions of observability and controllability (Ramadge and Wonham, 1982), and efficient use of system structure, such as is done with perturbation analysis (Ho, 1985; Suri, in press) or the likelihood ratio method (Glynn, 1986). All of these developments have shown that combining a dynamic systems view with the discrete event structure results in a fertile area for research (also see Ho, 1987).

• Development of more effective simulation tools. A manufacturing simulation model contains elements that include physical systems, control systems, management policies, external effects, etc. The use of interactive graphics and manufacturing terminology, from the start to the finish of a simulation task, can improve the effectiveness of simulation tools and make these techniques more usable. Building blocks are needed that will make such representations easy while retaining sufficient completeness.

• Efficient optimization of simulations. In the context of deterministic system optimization, a large number of software packages are available for system optimization. In contrast, no generally usable package exists for optimization of Monte Carlo simulations. Although theoretical papers abound, our experience is that the theory does not survive implementation! Much work needs to be done in this area, and there is room for substantial improvement over existing methods (Suri and Leung, 1987).

• Scheduling. This critical area of research will be discussed in a later section.

Implementation Phases

Although this paper is concerned principally with analysis, other important phases could properly be treated as a part of the manufacturing system life cycle, including procurement, installation, debugging and testing, and start-up. While each of these phases can benefit from analysis techniques such as project networks, Petri nets, software design, reliability, control theory, and many others, they are not as centrally affected by analysis as are the phases treated in more detail in this section. The bibliography in Suri (1985) and a relevant handbook (Charles Stark Draper Laboratory, 1984) provide further discussion of these phases.

Ongoing Operations Phase

The objectives of the ongoing operations phase can be divided into three main cate-

CORRESPONDENCE IN ACTIVITIES

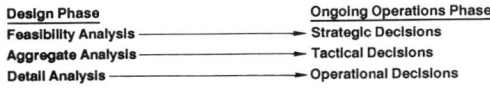

Design Phase	Ongoing Operations Phase
Feasibility Analysis ──────────►	Strategic Decisions
Aggregate Analysis ──────────►	Tactical Decisions
Detail Analysis ──────────►	Operational Decisions

FIGURE 2   Correspondence between activities in design phase and ongoing operation phase.

gories: strategic, tactical, and operational. These three differ primarily in terms of the length of the time frame in which they are to be considered. Strategic objectives extend over a period of years; tactical objectives, over several months; and operational objectives, from minutes to weeks.

The principal aim for the strategic phase is capacity planning—i.e., the development of long-range production plans and resource allocations—and system modifications—i.e., the changes that accompany new products, processes, or equipment. For the tactical phase, the principal aim is to develop aggregate production plans and aggregate resource allocations. For the operational phase, the main aims are the development of effective schedules, effective loading sequences, and job releases and the development of effective responses to disruptions due to breakdowns, non-availability of material, etc.

Specific issues, performance measures, and analysis techniques for the ongoing operations phase are not enumerated here. Instead we observe that there is a correspondence between activities in the three design phases (feasibility, aggregate, detail) and in the three ongoing operation phases (strategic, tactical, operational). That is to say, issues and techniques that apply to the feasibility stage during design also apply to the strategic decisions made during ongoing operations, and similarly for the other two sets of correspondences (see Figure 2). Thus, in designing a manufacturing system, analysis tools developed during the design phase can continue to be useful in all aspects of the operational phase. More positively stated,

since these tools can be important during the operation of a system, they *should* be developed during the design phase to generate more effective system designs.

Because of this correspondence, many of the research issues mentioned in previous sections apply here, too. We highlight a few critical ones:

• Scheduling. A particularly critical research issue is that of scheduling. Current OR-based approaches are either too simple in their assumptions or too complex to be readily solved or implemented. As noted by Milton Smith, Texas Technological University, at the 1984 ORSA/TIMS Conference on Flexible Manufacturing Systems, about 100 man-years had been expended by the OR community in solving minimum makespan scheduling problems, but he did not know of a single company that used minimum makespan to schedule shop floor operations (Gershwin et al., 1986). We need to have scheduling approaches that incorporate "real-world" considerations such as failures, shortages, schedule changes, and integration between corporate levels. In addition, these approaches should be simple to implement and fast to execute. There is a vast opportunity for new approaches or new ways of tackling this area. We should also keep in mind some of the lessons pointed out in earlier sections of this paper. Perhaps the solution lies not in trying to incorporate all the requirements of failures, shortages, etc., but in finding good ways to minimize such effects and then in finding simple scheduling approaches that work well under these new conditions. There is definitely an opportunity for completely new methods in this area.

• Real-time control of discrete event systems. Given the proliferation of shop floor monitoring systems and shop floor computers, it is natural that more will be expected of real-time control systems. Little is known about this area at the present time, except for a large body of heuristics. Recently there have been some successes using dynamic systems approaches (Akella et al., 1985). There is also a potential for applying many of the new approaches to discrete event systems mentioned earlier, in a real-time optimization mode (Suri and Leung, 1987), and for coupling these with traditional control-theoretic ideas (Gershwin et al., 1986).

• AI-based approaches. This is also an area where AI-based approaches may do well (Thesen and Lei, 1986). In fact, as mentioned earlier, AI-based methods, perhaps in combination with traditional approaches, may be suitable for solving some of the difficult problems identified in all the phases that have been discussed.

Obsolescence and Termination

The last phase in the life cycle is that of obsolescence and termination. Because there is less current interest in this phase than in the others, it is not covered in this paper. Suffice it to say that the elements of flexibility, efficiency, and responsiveness also play an important role in this phase of the life cycle.

DESIGN FOR ANALYSIS

The concept of design for analysis is a broad research issue that covers the entire spectrum of manufacturing system analysis. Briefly stated, design for analysis involves creating and working with designs that will be simple and easy to analyze. This approach may appear backwards to most engineers, who would view analysis as a tool that must serve the needs of the designers. They might also think that this approach would stifle the creativity and ability of good designers. Nevertheless, there are some strong advantages of design for analysis.

Precedents

There are precedents in the design field for changing the perceived user-resource relationships, and such changes can result in benefits. Up to the 1960s, it was taken for granted that a manufacturing designer knew best how to design products, and the job of the assembly engineer was to find the best way to assemble that product. The assembly engineer was considered to be a resource serving the designers. In the 1970s, however, the concept of design for assembly was introduced. This concept recognized that designing a product, while keeping in mind how it will be assembled, offered significant benefits, including lower cost and higher quality. This concept has since been broadened to include what is called design for manufacturability, thereby including all manufacturing processes. More recently, this approach has been expanded (Whitney et al., 1988, in this volume) to suggest that it can encompass strategic benefits as well. These developments have taken assembly and process planning and made them a part of the drivers that direct the user—the designer.

What Is Design for Analysis?

A central idea of design for analysis is that we should undertake the design of a product, or system, keeping in mind that the design will need to be analyzed. Therefore, if the design can be analyzed by using simpler and better-understood models, we may be able to model and analyze the concept faster, thereby allowing more time for examining various design alternatives, making sensitivity analyses, and asking "what-if" questions. The basic tenet is that this additional time will result in larger payoffs than the time spent in modeling and analyzing a more complex design.

Design for analysis, as an overall methodology, may result in other less obvious, but no less important, benefits. The best way to see these possibilities is through some examples.

Example from Solid Modeling

Voelcker (1988, in this volume) gives an example of the product and assembly description language (PADL) solid modeling system being used to represent the components in a Xerox copier. The simplest PADL system is the level 1.0 system (in terms of the number of different primitives it can represent). It was found that PADL-1.0 could fully represent 30 percent of the components in the copier. However, it was found that another 30 percent of the components could be redesigned so that they could be modeled by PADL-1.0 (Samuel et al., 1976). Voelcker now reports that a comparison of several of the redesigned components with their original counterparts indicated that the redesigned components were, in fact, superior from several points of view. This illustrates an unusual situation. Placing constraints on a designer may actually lead to a better design.

Example from Manufacturing System Design

Consider a hypothetical situation in which two teams independently undertake the design of a group technology cell to meet certain corporate manufacturing objectives. The design begins at the stage of selecting the products and equipment that should make up the cell. The design is to be done under typical time pressures. Team A wants to consider several different priority scheduling schemes, operator policies, and lot-splitting techniques to optimize the performance of the cell. The team decides to put together a simulation model to allow all of these parameters to be studied. Team B decides to consider simple performance policies and finds that a spreadsheet package along with a simple analytic modeling

tool (e.g., Suri and Diehl, 1987) are sufficient to study the alternatives. It is our claim, not as yet based on scientific research but only on personal observations of industrial operations, that Team B is likely to arrive at a strategically more effective design. As a result of the time pressure, Team A is likely to spend a good deal of time developing the complex model and not enough time on exploring the alternatives. Team B will start exploring alternatives early in the process and will quickly identify some of the critical parameters to be examined. It will have an opportunity for several iterations of designs, and it will have explored a wide range of alternatives. In this hypothetical example, although Team B was constrained to look at a smaller class of solutions, with the analytic tools that it chose to use, it was actually able to explore more alternatives under the "real-world" pressure. So the design for analysis principle would encourage the elimination of certain designs that are too hard to analyze in favor of designs that are simpler to analyze.

What Design for Analysis Is Not

Since the concept of design for analysis may seem upsidedown to many engineers, it may be useful to further clarify it:

• Design for analysis does not mean distorting reality to fit our models. It does mean asking, "Should we change what we have chosen to define as reality to fit our model?" and "What might be the benefits of this change?" It forces us to think about the impact of this change and not take reality to be determined by a given situation that must then be modeled.

• Design for analysis does not mean we should stifle development of modeling technology. We should definitely be pushing the frontiers of what we can do with analysis and modeling. It does, however, focus on using today's analysis technology most effectively.

• Design for analysis is not just a different way of saying "design for simplicity." Designing for simplicity is an important goal that many industries are recognizing today. Design for manufacturability is one step toward meeting that goal. Design for analysis is also a step toward that goal, in that it offers specific approaches and yardsticks, through the available analysis tools, by which to learn whether simplicity is being achieved.

Research Potential

There is more depth to this methodology than may be evident from the foregoing brief description, and there is scope for novel research on this concept. Design for analysis can lead to

• Simplicity. In science this is usually coupled with elegance of solution and concepts.
• Robustness. The resulting design works better over a wide range of conditions.
• Responsiveness. Designs and manufactured products can be created faster and more accurately.
• Simpler operations. Systems designed using this method will be easier to manage, operate, and change.
• Strategic focus. Teams working with this method will find themselves forced to look at strategic issues and not just tactical responses.

We do not claim that all of these results will follow in all situations, but we do believe that they can follow in a number of nontrivial instances. The research issues that follow from this concept can be stated as follows:

• Establish a number of instances, from different disciplines, where design for analysis leads to recognizable benefits (e.g., the Voelcker case study described earlier) and at the same time identify counterexamples —i.e., instances where its use is counterproductive.
• Develop an understanding, perhaps leading to a theory, of why this method works and when it can be expected to work; for example, why did it produce a better component for the copier?
• Determine the building blocks that will make systems that are simple to design and analyze, and it is hoped as a consequence, simple to implement, operate, change, and manage.

CONCLUSION

There is a need for good "traditional" analysis approaches to be used in conjunction with recent manufacturing methods— JIT, AI, etc.—and whole new areas for research are just opening up. However, it is essential that we work closely with real systems and that we keep abreast of recent developments. On a more specific note, we offer the following concrete suggestions of what this research community can do.

Individuals must make a commitment to get to know the area. They should work closely with industry. One could set a personal goal to work a given number of days in a factory each year, or to get a certain amount of industrial funding, or to be involved in industrial projects. Individuals should also stay up to date on technologies, approaches, and philosophies in this field. Only after taking these basic steps should one look into the research areas in this field.

The community—professional societies, journals, and research establishments— must also encourage developments in positive directions. There should be an emphasis on, and rewards for, good applications. There should be better mechanisms and incentives for joint industry-university projects. We must improve the exchange of ideas among theoreticians and practitioners, perhaps through special workshops or publications. Finally, we must reexamine our attitudes and journal acceptance processes so that emerging new ideas are better nourished, rather than simply publishing incremental contributions on safe, accepted topics. Our long-term health depends on continued innovation, not just refinement.

The aim of this paper is not to present one person's view but to encourage the whole community to think more about these issues. If further introspection and dialogue increases the viability of the community, the resulting developments can be very beneficial.

Acknowledgment

This work was partially supported by the National Science Foundation under Grant No. DMC-8717093.

REFERENCES

Akella, R., Y. Choong, and S. B. Gershwin. 1985. Real time production scheduling of an automated cardline. Pp. 403–425 in Flexible Manufacturing Systems: Operations Research Models and Applications, K. E. Stecke and R. Suri, eds. Basel, Switzerland: J. C. Baltzer AG.

Charles Stark Draper Laboratory. 1984. Flexible Manufacturing Systems Handbook. Park Ridge, N.J.: Noyes Publications.

Cohen, G., D. Dubois, J. P. Quadrat, and M. Viot. 1985. A linear-system-theoretic view of discrete event processes and its use for performance evaluation in manufacturing. IEEE Transactions Automatic Control AC-30:210–220.

Fox, M. S., and S. F. Smith. 1984. A knowledge-based system for factory scheduling. Expert Systems 1:25–49.

Gershwin, S. B., R. R. Hildebrant, S. K. Mitter, and R. Suri. 1986. A control perspective on recent trends in manufacturing systems. Control Systems Magazine 6(2):3–15.

Glynn, P. W. 1986. Stochastic approximation for Monte Carlo optimization. Pp. 356–365 in Proceedings of the 1986 Winter Simulation Conference. New York: Institute of Electrical and Electronics Engineers.

Haider, S. W., and J. Banks. 1986. Simulation software products for analyzing manufacturing systems. Industrial Engineering 18:98–103. (Also see errata and corrected figures in Industrial Engineering 18:86–87.)

Haider, S. W., D. G. Noller, and T. B. Robey. 1986. Experiences with analytic and simulation modeling for a factory of the future project at IBM. Pp. 641–648 in Proceedings of the Winter Simulation Conference.

Hall, W. 1983. Zero Inventories. Homewood, Ill.: Dow Jones-Irwin.

Hayes, R. H., and S. C. Wheelwright. 1984. Restoring Our Competitive Edge: Competing Through Manufacturing. New York: Wiley.

Ho, Y. C. 1985. A survey of the perturbation analysis of discrete event dynamic systems. Pp. 393–402 in Flexible Manufacturing Systems: Operations Research Models and Applications, K. E. Stecke and R. Suri, eds. Basel, Switzerland: J. C. Baltzer AG.

Ho, Y. C. 1987. Performance evaluation and perturbation analysis of discrete event dynamic systems: Perspectives and open problems. IEEE Transactions on Automatic Control AC-32(7):563–572.

Lundrigan, R. 1986. What is this thing called OPT? Productivity and Inventory Management 27(2):2–12.

Meleton, M. P., Jr. 1986. OPT—Fantasy or breakthrough? Productivity and Inventory Management 27(2):13–21.

Ramadge, P. J., and W. M. Wonham. 1982. Supervision of discrete event processes. Pp. 1228–1229 in Proceedings of the IEEE Conference on Decision and Control.

Samuel, N. M., A. A. G. Requicha, and S. A. Elkind. 1976. Methodology and Results of an Industrial Part Survey. Report TM-21, Production Automation Project, College of Engineering, University of Rochester.

Schonberger, R. J. 1982. Japanese Manufacturing Techniques. New York: The Free Press.

Skinner, W. 1985. Manufacturing: The Formidable Competitive Weapon. New York: Wiley.

Suri, R. 1985. Quantitative techniques for robotic systems analysis. Pp. 605–638 in Handbook of Industrial Robotics, S. Y. Nof, ed. New York: Wiley.

Suri, R. In press. Infinitesimal perturbation analysis for general discrete event systems. Journal of the ACM.

Suri, R., and S. DeTreville. 1986. Getting from "just-in-case" to "just-in-time": Insights from a simple model. Journal of Operations Management 6(3):295–304.

Suri, R., and G. W. Diehl. 1987. Rough-cut modeling: An alternative to simulation. CIM Review 3:25–32.

Suri, R., and Y. T. Leung. 1987. Single Run Optimization of Discrete Event Simulations: An Empirical Study Using the M/M/1 Queue. Technical Report 87-3, Department of Industrial Engineering, University of Wisconsin—Madison.

Thesen, A., and L. Lei. 1986. An expert system for

scheduling robots in a flexible electroplating system with dynamically changing workloads. Pp. 555–566 in Proceedings of the Second ORSA/TIMS Conference, K. E. Stecke and R. Suri, eds. Amsterdam: Elsevier.

U.S. Congress, Office of Technology Assessment. 1984. Computerized Manufacturing Automation: Employment, Education, and the Workplace. Office of Technology Assessment, OTA-CIT-235, Washington, D.C.

Voelcker, H. B. 1988. Modeling in the design process. In Design and Analysis of Integrated Manufacturing Systems, W. Dale Compton, ed. Washington, D.C.: National Academy Press.

Whitney, D. E., J. L. Nevins, T. L. De Fazio, R. E. Gustavson, R. W. Metzinger, J. M. Rourke, and D. S. Seltzer. 1988. The strategic approach to product design. In Design and Analysis of Integrated Manufacturing Systems, W. Dale Compton, ed. Washington, D.C.: National Academy Press.

SIMULATION IN DESIGNING
AND SCHEDULING
MANUFACTURING SYSTEMS

F. Hank Grant

ABSTRACT   As manufacturing companies strive to achieve increased efficiencies, they must make effective use of technology. Simulation is an important tool in accomplishing this.

The use of simulation for scheduling and control of production systems is a natural outgrowth of its application for the design of systems. Simulation, when used for production scheduling, is a useful vehicle for providing the discipline necessary for effective production management of the factory floor.

This paper discusses the applications that provide and support the use of simulation for the design and operation of integrated manufacturing systems. A discussion is given of the new technology that makes simulation available for production scheduling. The differing objectives of the production system scheduler and the production system designer are discussed. Important research topics in simulation are also identified and discussed.

INTRODUCTION

Manufacturing companies have a pressing need to understand new technology and the potential for its use in a rapidly changing environment. Although simulation has been used traditionally to help explore the ramifications of new technology for manufacturing systems, many companies now require that simulation studies of proposed manufacturing facilities be performed before a final decision is made on implementation of either new or current technology. Simulation can provide insight into issues that are not apparent or are counterintuitive. Many software tools now available provide engineering support for system design projects.

The use of simulation for scheduling and control of production systems is a natural outgrowth of its application in design. The objective in designing a manufacturing facility is to obtain knowledge sufficient to make capital commitment decisions that satisfy production objectives, and these same characteristics are also important in controlling production. Simulation, when used for production scheduling, provides a vehicle for achieving the discipline necessary for effective production management on the factory floor, thereby helping to achieve productivity goals through efficient, effective use of new, integrated technology. New computer technology, such as powerful workstations, coupled with advances in easy-to-use software systems, makes simulation a useful tool for the production scheduler.

Simulation is an analysis tool that can be applied effectively to a variety of shop floor design and real-time shop floor control problems. Simulation can support longer-

134

term system design evaluations of resource requirements, equipment needs, inventory buffer sizes and availabilities, and sensitivity analysis of a variety of product demand and equipment performance probabilities. Simulation can also support shorter-term decisions involving equipment scheduling, shop order release, and work order scheduling.

Historically, simulation techniques have been highly successful and used extensively for the planning and analysis of current operations or proposed designs. Although on-line simulation analyses are feasible and can be cost-effective, applications of simulation in this mode have been few. On-line applications of simulation for real-time shop floor control purposes can be applied effectively only if the data supplied to the simulation models are accurate, organized, and timely. In the past these constraints have proved difficult to overcome.

More recently, major advances have been made in data-base technologies. These improvements make the access and manipulation of data for real-time factory control a reality. A modeler can now construct a simulation model of an operating unit and supply this model with accurate and timely data describing the performance of machines and the operators and the expected demands on the manufacturing system.

Other data processing improvements that facilitate the use of simulation for on-line analysis relate to computational capability and the graphic display of results. Managers have historically disliked having to wait a long time for analysis. With the computing capabilities and graphic constructs that are now available, managers not only can get quick response for analysis of work order scheduling or work order release, they can also receive the information in more easily understood form.

This paper discusses how these latest developments will support the use of simulation for designing and controlling production systems. It also identifies and discusses the important research areas and the challenges in effectively applying this technology.

SIMULATION TECHNOLOGY FOR DESIGN

Simulation is the process of creating a representation or model of the operation of a system on a digital computer. This model is created by providing to the computer a description of the physical components of the system and the logic associated with the operation of the system. The model of the system under study is put together using a simulation language. This language gives structure to the model-building process by providing special constructs that relate to the system under study. For example, many languages provide constructs to represent components like queues and servers. Other more system-specific languages provide constructs to represent machines, parts, and process plans.

Many different simulation languages are available. These include general-purpose languages, such as SLAM II, SIMSCRIPT II.5, and GPSS, and more specialized languages oriented toward manufacturing, such as MAP/1, SPEED, and MAST (Haider and Banks, 1986; Miner and Rolston, 1983; Pritsker, 1986).

Once built, the simulation model serves as a laboratory in which various design alternatives can be tested and compared. By running the model on a computer, the actions of the proposed system are represented in detail, permitting inferences to be drawn about overall system performance. These inferences are made on the basis of numerous performance measures provided by the simulation, such as machine utilization, in-process inventories, part waiting times, and throughput. Through this process of experimentation, the best overall system design is selected.

Several features of simulation make it particularly advantageous for the design of manufacturing systems (Musselman, 1984). These features include physical and control system balance, system flexibility, random system behavior, and animation.

Physical and Control System Balance

Proper balance among the physical and the logical and control components of a manufacturing system must be maintained during system design. Should the design of one component be emphasized disproportionately, overall system performance could suffer. Designing a system with proportional importance given to each component is possible through the total systems perspective that simulation provides.

For example, in modeling a flexible manufacturing system (FMS), the system can be viewed as being composed of two interrelated subsystems: a physical subsystem and a control subsystem. The physical subsystem, which transports, stores, and processes parts, consists of programmable machines, material handling equipment, and in-process storage facilities. The control subsystem, which selects, sets priorities, routes parts, and controls traffic flow, consists of situation-dependent logic to coordinate part interchange in the physical subsystem. These two subsystems must work in concert. It is important, therefore, to consider the interaction between these two subsystems during the design. Isolating, for analysis purposes, one subsystem from the other or giving one preferential treatment will most likely result in misleading conclusions.

A more complete evaluation, with appropriate consideration being given to the interplay between these two subsystems, is possible with simulation. In contrast to strictly analytical approaches, simulation includes the system's operating and control strategies as an integral part of the model. As parts move through the various operations, the model processes them according to each machine's operating characteristics and routes them according to the system's situation-dependent control logic. Simulation provides a rich environment for the design of physical as well as logical and control systems.

System Flexibility

The increasing popularity of simulation is due, in part, to its ability to represent various levels of detail. Analytical formulations, while offering closed-form solutions, tend to be restrictive, since details must often be neglected in order to accommodate the formulation. Simulation, on the other hand, can provide as much or as little detail as the analyst wants. All relevant system characteristics can be taken into account, such as processing time variability, equipment reliability, fixture restrictions, in-process storage requirements, complex routing decisions, operating policies, and scheduling constraints. Simplifying assumptions, such as reduced decision models, are not required. The result is a flexible experimental setting in which to test alternative design strategies where the analyst has control over the details and assumptions that are included.

Random System Behavior

The operational behavior of manufacturing systems can be quite dynamic over time. That is, the interaction of various system components can trigger unexpected behavior that may occur only infrequently. The system can exhibit, for example, significant variation in demand for resources, depending on the particular interactions of its components. Although this dynamic behavior tends to increase the complexity of the design, and often the model, knowledge and understanding of this behavior can lead to significant design efficiencies. Simulation provides a means of understanding system dynamics and learning about components of the system that may interact and behave in a counterintuitive manner.

Proper accounting of random variability of system parameters helps ensure that a proposed design will exhibit stable performance in a variety of configurations. Ironically, instability can actually be promoted when certain sections of a system are selectively overdesigned to guard against processing variations. As a result, there can be unforeseen weak links in the system. Through systematic experimentation with a model, weak links can be revealed and actions taken to correct them.

An example of this use of simulation is in the study of fixture load stations. To examine the effectiveness of a particular configuration for a fixture load station, the part arrival sequences and their demand on the station must be understood. Unfortunately, given the variety of part types that could be simultaneously processed and the processing variations that could occur for each part type, the number of different part arrival sequences is extremely large. Examining all possibilities in great depth is unreasonable. However, rerunning a representative, randomly selected set of part mixtures in the simulation model can provide insight into the station's performance. If the design of this station were found to be deficient, it could be modified and then tested against a broader range of part arrival sequences. This ability to deal with random behavior with the models and to create representative samples that stress the system's capability is a major benefit of simulation.

Animation

One aspect of simulation technology that has made considerable progress in recent years is graphical animation. Besides the usual plots of system performance, it is now possible to animate a manufacturing system operation in great detail (Grant and Weiner, 1986). By bringing the model of the physical subsystem to life graphically, the operation of the control subsystem can be studied in action. This provides a detailed understanding of the implications associated with the system's control policies. Subtle errors can be identified and corrected before the system becomes operational.

SIMULATION TECHNOLOGY FOR SCHEDULING AND CONTROL

The same simulation technology that has been used traditionally in system design can be used as the kernel for day-to-day production scheduling applications. As an introduction to the application of simulation to detailed production scheduling, traditional tools, the limitations of these tools, and the ways that simulation technology can address these limitations are discussed.

Traditional Scheduling Methods and Their Limitations

Shop floor scheduling is an important task in managing a production system. It entails complex decisions that affect such objectives as meeting delivery due dates and maintaining a desired level of inventory. Although it is not possible to consider all of the variables that determine the effectiveness of a particular schedule, major productivity improvements can be realized by making the production scheduling process more effective.

The quality of a production schedule involves many, sometimes conflicting, objectives. Whereas maximizing throughput is certainly one important consideration, an ideal schedule will also have the following characteristics:

• Delivery due dates will be met.
• Inventory costs will be maintained at acceptable levels.
• Equipment, personnel, and other limited resources will be well utilized and have balanced workloads.
• Adaptations can be made quickly in

the case of an unexpected event, such as equipment failure or raw material shortage.

Since it is difficult to "optimize" a schedule over all these characteristics, one or two characteristics are often chosen, depending on current production objectives. Generally, trade-offs must be made to reach a balance between these more limited objectives.

Production scheduling is done in many ways in industry. Probably the most common methods of scheduling are purely manual techniques. In the most straightforward form, an expert such as the department foreman or the machine operator selects the next job from those waiting in front of the machine. The criteria used in this circumstance often reflect the measures by which the scheduler is evaluated and may not reflect overall business objectives. Job status control boards are also used to lay out schedules visually.

A more analytical approach to scheduling is sequencing by dispatching rules. This method uses rules that set priorities for the jobs waiting for processing. Research has demonstrated that rules such as the weighted shortest processing time rule can generate reasonable schedules. The effectiveness of the schedule may vary widely, depending on the particular rule selected, the type of production facility, and the mix of jobs to be produced. It is impossible to predict which dispatching rules will work best in most manufacturing systems by traditional methods. They are also limited in the scope of what they consider and are often hard to implement on the shop floor in a cost-effective manner.

Material requirements planning (MRP) was one of the earliest computerized techniques for factory management. In its earliest form it generated unconstrained production schedules, which were based on a bill of materials and estimated production time requirements. Manufacturing re-source planning (MRP-II) expanded the scope of MRP to consider many other facets of production facility management. In addition to sophisticated factory accounting capabilities, modules for capacity planning and shop floor data collection were also provided. These techniques are effective for longer-term scheduling and order launching, but they lack the detail necessary for effective day-to-day production scheduling.

Recently, there has been emphasis on more sophisticated definitive capacity planning tools as a means of generating production schedules. These methods determine the expected critical resource and then schedule forward and backward around that resource. At present, these techniques do well on a global level but are inadequate when highly interactive components are present and frequent changes are needed. The computational time required to generate schedules is usually large, with interactive execution being impractical. Also, the critical, or bottleneck, resources tend to change, based on production demand.

As an alternative to these approaches, simulation-based scheduling can provide an effective tool for shop floor scheduling while requiring few assumptions. The schedules generated are based on an accurate, realistic model of the production facility.

Simulation-Based Scheduling

Simulation practitioners are familiar with the ability of simulation models to predict system behavior in great detail. It is a natural extension to attempt to apply simulation on a day-to-day basis to predict schedule performance and resolve problems before they occur. Simulation is well suited to this type of analysis and, with proper support, can be used successfully to create and evaluate production schedules. The simulation model provides a computer replica of the department in the factory. The model plays through the schedule and provides performance information. Event trace information can provide the details of a feasible schedule, given the constraints specified in the model. The detailed interaction of various production limitations can be included at any level of detail.

The dynamic interactions between resources can be captured and analyzed with simulation. For example, a material handling vehicle may deliver product to several machining stations. Even though this vehicle may have an expected total workload requiring only 50 percent of its available time, two stations that need material at the same time will cause the system to perform differently than it would if there were no conflict for the vehicle's services. Simulation can be used to evaluate the effect of these conditions on schedules and to resolve the conflict.

In another case, the scheduler may have many options for selecting the transfer batch size in a production order. The entire order may be processed at each operation, or the order may be split into two or more separate loads and allowed to move independently through the operation. Simulation would allow the scheduler to contrast alternatives and select the one that provides the best performance.

Because this analysis is performed on a computer, many alternatives can be explored with relatively little expense or risk. Simulation models can represent a large variety of the factors critical to a manufacturing system's performance in great detail. The influence of tooling, personnel, and other resources can be evaluated. Capacity changes, such as a machine breakdown or scheduled maintenance, may also be included in the model.

Most simulation languages have the modeling features necessary to represent production systems. But there are many additional needs in scheduling that are not typically provided. To be effectively used for scheduling, simulation models must provide reports that can be readily understood and used by the production sched-

uler. Preferably, reports should be provided that serve as actual job performance schedules for equipment and personnel and are consistent in format. Also, a strong user interface is needed to support the scheduler and to allow schedules and other reports to be created easily. The scheduler should not have to interact with the details of a simulation model but still should be able to use it easily. Most simulation languages do not provide these features. The requirements for simulation in a scheduling and control mode are discussed in more detail in a later section.

Traditionally, large simulation models of production facilities have been too expensive to build and too cumbersome for use on a daily basis. The modeling process has required a highly trained and experienced analyst with a solid understanding of both the simulation language and the system under study. Even after the model is completed, execution of detailed models could take too long to be useful in production scheduling. This problem has been addressed in part by the advance of computer technology. Workstations that provide mainframe-like capabilities are moving rapidly to the factory floor. With the decrease in price of local area networks for factories, and their increasing use, computers that focus on scheduling will proliferate. Also, more advanced simulation languages permit faster execution and more efficient model development. These languages are tailored to the simulation of manufacturing systems. The following overview of an existing tool illustrates how simulation can be used in this mode.

One Available Scheduling Tool

Currently there are few commercially available products that provide tools for applying simulation to production scheduling. One product that addresses scheduling is called FACTOR™ (Grant, 1986, 1987; Pritsker et al., 1986). It is designed to be highly interactive and used by shop floor scheduling supervisors.

FACTOR™ supports two types of users in providing a shop floor scheduling system. The primary user is the factory floor production supervisor, who uses it to generate production schedules. The second user is the individual responsible for modeling the factory and developing the data that characterize the manufacturing system and the products that are produced in it. The model development activity typically occurs only once in the installation of FACTOR™. Modification of the FACTOR™ model is easily accomplished to incorporate changes in the production department after it has been installed.

FACTOR™ can be integrated with the production management software and plant data to provide a useful tool for the production scheduler. One important feature that FACTOR™ provides is an interface to the MRP system to extract information related to a master schedule. It also interfaces with the shop floor status system to acquire production data. In FACTOR™, this is typically a straightforward interface to the particular MRP system that is in use. This interface collects the current list of orders to be processed at the production department and automatically loads them into the FACTOR™ system. A similar interface is provided to collect information regarding the current location and status of released orders from the shop floor data collection system. This information is also loaded into FACTOR™ to initialize the scheduling system automatically to start at current shop floor status. These data are stored in a local data base that may be accessed and edited by the production supervisor.

Once the scheduling supervisor has gathered the current production requirements, FACTOR™ is used to simulate production and identify potential problems. The supervisor can select from a large variety of ranking priority codes to sequence jobs through the production system. These codes include familiar priority measures such as due date and shortest processing time. In addition, more complex ranking priorities are provided, such as estimated remaining processing time based on past performance as well as a critical or tactical ratio. The critical ratio measure is the ratio of the scheduled remaining production time to the estimated production time. The tactical ratio is the inverse of the critical ratio. Additional priority schemes are available based on production costs and other system status variables.

Tools such as FACTOR™ make many contributions, including implementation of technology necessary to schedule the factory floor. Another is the development of the framework necessary for further developments in scheduling technology.

The next section describes a set of requirements to support the application of simulation in production scheduling and compare those with requirements in design application.

THE FOCUS IN SCHEDULING AND CONTROL VERSUS DESIGN

There are significant differences when applying simulation for production system design as opposed to production scheduling. Table 1 summarizes these differences.

Effectiveness of the User Interface

An issue of primary importance is the effectiveness of the user interface. Simulation for the design environment uses a language-based interface, and the end user typically builds the model. Model management tools, such as the TESS system supplement (Standridge and Pritsker, 1987), have the ability to provide model-building tools that interface through a graphics device. Their focus is primarily on building the model and managing the simulation analysis. Scheduling and control applications have a need for significantly more powerful

TABLE 1 Issues and Requirements for Simulation Technology Applied to Scheduling and Control versus Design

Scheduling and Control	Design
1. Strong user interface is needed to define the production replica and to generate production schedules.	1. Language-based interface is used to build the design model and generate performance reports for various design alternatives.
2. Implemented set of algorithms for sequencing production orders is used.	2. User-designed and -coded algorithms associated with queue-ranking procedures are used.
3. Execution of the system is interactive.	3. Execution of the system is typically in a batch mode.
4. An interface is provided to external data sources to integrate the simulation-based scheduling tool into base management systems.	4. Any interfaced external data must be explicitly created by the user, and most design-mode simulation languages are not created to support an interface to external data.
5. Production schedule reports are explicitly produced for the factory floor.	5. Standard performance reports are provided for system performance measures, but any specific reports must be user-coded.
6. All input and output is stored in a data base that can be interfaced to other systems.	6. Data-driven design-level tools typically use data stored in a flat file, but input data and output data are sometimes stored in a data base, as is provided by a simulation study management system.
7. Internal design of the software is oriented toward fast execution to respond to the needs of the production scheduler.	7. A general-purpose language addressing a wide variety of system and analysis objectives will require moderate execution times.
8. Orientation of the model is toward production planning and setting production objectives.	8. Orientation of the model is toward broader design issues, including randomness as a standard component.
9. Technology is an application generator that supports two users: the modeler and the scheduler.	9. Technology is a modeling tool and focuses on design models for the modeler-analyst.

tools to support the production scheduler. Such tools are needed since production schedulers tend to be experts in their particular manufacturing operations but not in simulation or other computer technology. An easy-to-use interface must be provided to support the generation of production schedules as well as the need to manage the implementation of those schedules on the factory floor. The user interface is also an excellent application for expert system technology. Expert system tools can be used to analyze the need to reschedule based on progress toward expected performance. They can also identify problems in production schedules developed for alternative

scenarios and present them in an easy-to-interpret format. They can extend the scheduler's experience and greatly ease the data analysis requirements.

Support Features for Production Scheduling Models

The second issue is concerned with support features required for the production scheduling models. In the design mode, the user typically develops and codes the various algorithms used in the model. In the scheduling and control mode, the user requires an implemented set of algorithms that can be easily accessed to support scheduling applications. Further, the scheduler will typically access more than a single algorithm in any scheduling exercise. From these alternative scenarios, the alternative generating the best performance will be selected for implementation.

Execution of the System

In the design mode, execution of the model is typically done in a batch process without much interaction. The exception to this procedure occurs when animation is used to view the simulation where the capability exists to interrupt the animation and review and modify certain system parameters. Execution of the model in the scheduling and control mode must be interactive. The scheduler must have the ability to interact with both the management system and the scheduling tools to review the data provided and make production management decisions.

Data Needs for Production Models

Extensive external data requirements exist in scheduling and control applications. It is necessary to provide an interface that allows the scheduling supervisor to integrate the scheduling control system easily into the other management tools that are used in the production facility. The interface must be able to download the data easily whenever required into the scheduling model and to develop the necessary schedules. In the design mode, the user is focused on specific design tasks and has little need for a day-to-day interface with other production data. The data requirements are addressed by gathering performance data to assess the impact of the design on productivity goals. The interface to external data for scheduling and control applications must be very general as well. Each factory has different data storage characteristics and capabilities, and to be

effective in this area, a tool must be able to address a variety of applications.

Reporting Requirements

Reporting needs in the design mode are focused on system performance measures and are usually tied to simulation constructs. Special-purpose languages are alleviating this issue in providing reports in the context of system-specific issues, such as machine performance or operator utilization. In the scheduling and control mode, reports must be provided in a form explicitly applicable to the factory floor situation. This includes not only performance reports but also detailed production schedules that the operators can follow and execute. A report generator is also required to provide additional presentation flexibility.

Storage of Data in a Data Base

Although many design tools have adopted the philosophy that input and output data should be stored in a data base, several others provide input data in the form of a flat file. They also store output data in a flat file. In a schedule and control application, all input and output data must be stored in a data base. The user has the need to query the output data base interactively to evaluate various performance issues as well as to compare alternative scenarios. That same need exists in the design mode, but it is even more critical in production scheduling because of the time constraints imposed on the production scheduler.

Ability to Address a Variety of System and Analytic Objectives

The requirement to address a wide variety of objectives typically imposes constraints on execution speed. Given the breadth of application design, there are bounds on the execution efficiencies that can be achieved in these tools if they are to execute quickly. In a scheduling and control mode, the application is focused on a specific objective—to generate achievable production plans in the shortest possible time. Therefore, the internal design of the simulation software kernel can be oriented toward fast execution in response to the needs of the production scheduler. Issues such as stochastic variables and interpretive languages, which are important in a design mode, may be eliminated entirely in the production scheduling applications.

Breadth of Demands on the Models

Design issues are typically relatively broad. The user is focused on exploring alternative designs in the production system and evaluating the response of the system to various stresses in production demands. In scheduling and control, the focus is to develop a replica of what exists on the factory floor in order to plan production. The approach is not to run the simulation and see what happens, which tends to be the objective of design, but rather to run the model and produce an achievable plan that the factory floor can execute. This yields a significantly different model from that in the design mode. These models may be more focused and often require less detail and scope than those required for design. They often require, however, significant detail in the representation of the decision processes that occur on the factory floor. These processes tend to require very explicit rules that emphasize the characterization of existing procedures.

Users of Simulation

Simulation technology for scheduling production systems typically addresses a different group of users than simulation technology applied in the design mode. When used in the scheduling and control mode, the technology is actually an application generator that the production modeler uses to create a scheduling system specific to the production environment for use by production schedulers. The modeler can be an engineering group or a data processing group that is closely tied to the manufacturing environment. The scheduler, of course, is someone with production scheduling authority on the factory floor. In the design mode, the technology tends to focus on the needs of the system designer, who is typically someone from an engineering group. The user's objectives are to explore various alternatives for meeting production goals and objectives and to justify the cost-effectiveness of the proposed system.

RESEARCH OPPORTUNITIES

Simulation has been used for the design of industrial systems for many years. Its use as a scheduling and control tool is more recent. Significant opportunities for further research exist in both applications. Although the design mode is certainly better developed, significant problems still exist. The application of simulation technology in scheduling and control is an extremely fertile area. The following sections describe the more important research to be addressed and provide additional details on the opportunities and challenges of each research area.

Use and Integration of Artificial Intelligence Tools in Simulation

Artificial intelligence is an extremely broad area and has many possible applications in simulation. The most attractive is expert system technology. Research has demonstrated the feasibility of developing simulation in an expert system framework and using standard expert system technology to generate the simulation executive, models, and outputs (Reddy and Fox, 1982). A major difficulty with this approach has been the execution speed of the programs.

A primary opportunity for the application of expert system tools is in support of output analysis. Simulation requires significant effort on the part of the user to extract and condense the large amount of information that is produced. Expert system technology could drastically condense the output data and produce a knowledge base with which the user could interact to assess the effectiveness of a design.

Expert system tools should also be able to support the design task as the model is developed. Tools are required for developing the model, characterizing components of the model, and relating them to system objectives and analysis output. A related technology is the application of expert system tools to computer programming.

Significant research opportunities also exist for the application of expert system technology in scheduling and control. First, expert system technology can be applied to capture expertise on the factory floor and integrate that information with the model. This application would include rules for decision making regarding characteristics of production sequencing. Such rules could become part of the scheduling system. The integration of expert system technology with simulation models should be relatively straightforward where the technical challenges will be to implement the integration in such a way that simulation execution performance is not degraded.

A second application area is concerned with output analysis. To make decisions, production schedulers must digest the vast quantities of data produced by a scheduling system. Expert systems could perform data reductions as well as analyze the output. They could make decisions to generate alternative scenarios and select the best. This application of expert system technology would also be of use in automated environments, where the user of the scheduling module might be a cell controller, or another CPU. In this circumstance, the need is to be able to reschedule and evaluate the results of the simulation scheduling application without human intervention. If the analysis procedures could be captured in an expert system, the decisions could be made automatically.

The third area for the application of expert system technology is concerned with analyzing performance to plan and determine the need to reschedule the system. Production systems will always experience perturbations around a plan. The support needs center on determining the significance of those perturbations, defining the need to reschedule, and determining different methods of responding to a rescheduling requirement.

Real-Time Data Collection and the Interface with Models

Factory floor data collection systems have existed for several years and are available in a variety of forms. They collect a wide variety of data, from production order tracking to detailed machine status. They have had varying degrees of success, primarily because of a lack of significant application for the data collected other than simple status reporting. Typically, by the time the information was collected and reported, it was too old to be of use. New developments in manufacturing, particularly in automated production systems, can provide the means for maintaining current data about the system. As the price/performance ratio continues to improve in computer technology, the use of effective data collection systems will increase in manual operations as well as automated systems.

The primary application for factory status data is in scheduling and control applications of simulation. Design activities are typically much longer term and, although some of the performance data collected would be of use, a real-time interface is not necessary.

Real-time data collection and its interface with scheduling and control applications offer interesting research opportunities in many areas. The scheduling and control systems obviously require good status information to generate production schedules effectively. A real-time interface to the data collection system would permit the scheduling system to access those data whenever necessary. In fact, the simulation model could be run in parallel with the physical system to determine the rescheduling needs dynamically. Real-time collection of information on performance could also drive certain parameters in the simulation model to increase its realism while still providing effective projections for control of the production system.

Data between the physical system and the scheduling models also flow to the physical system. Simulation models can generate a detailed production plan, and that plan can be implemented through the flow of data from the model to the production system. This can be implemented at a variety of levels of detail, starting with an agenda prescribing specific manufacturing operations for the various components of the production cell. More detailed scheduling models could actually drive physical equipment and serve as an important part of the cell control software.

Data Integration and Distribution

Data integration is a significant issue in the application of simulation in production systems. The sources for data that drive the production system are typically distributed throughout the manufacturing control system and may require significant time and effort to organize and use effectively. A research opportunity exists in developing effective methods for describing the integration of data across all manufacturing operations and also for providing tools for effectively transferring the data from component to component. In a design mode, data are needed to characterize existing production policies and performance characteristics and integrate those into the model to accurately predict changes in production characteristics and evaluate the design.

When used in a scheduling and control mode, data integration from distributed data sources is critical. The data needed to support effective simulation models for scheduling and control range from the list of orders from the MRP system down through shop floor status information. Machine characteristics, machine status, process plans, part descriptions, etc., are also required. In most corporate systems, this information lies in various sets of data bases. Current applications use a standard ASCII file to transfer data. Additional research is needed to define how the data may be more effectively distributed and accessed. It is clear that the success of additional production management tools using simulation technology depends on their ability to be easily integrated into existing components.

Interaction and Integration of Simulation with Automated Systems

Automated systems are being implemented throughout industry. Research opportunities for implementation of simulation in automated systems exist primarily in scheduling and control applications. Cell control software development efforts can be greatly reduced if the technology to schedule the cell and react to various problems that occur in production are included as a part of the control software development. It has been shown that this function can be successfully implemented as a separate module. This allows the cell control software to be relatively simple and focus on hardware control issues as opposed to scheduling issues. Implementation of scheduling within the cell controllers also tends to create a significant overhead burden on the control hardware, since it requires continuous iteration to regenerate the schedule.

Production scheduling systems, such as the one described earlier in this paper, can be interfaced with automated systems to provide scheduling tools to generate detailed production schedules effectively and provide a significant analysis capability as well. Cell control software can invoke the production scheduling system to run several scenarios and generate a "best schedule" for implementation in the automated cell. This can often be done without human intervention when various scenarios for analysis have been set up ahead of time.

Several significant research questions remain. One of the most important is concerned with data integration in the automated system. Data integration requires solving the problem of effectively interfacing the automated system with the scheduling and control models to generate detailed production schedules efficiently. This process may require frequent access to data by the scheduling systems as production is monitored and the need to reschedule arises. Another issue is concerned with the analysis requirements that must be addressed in providing scheduling systems for automated cells. These analytic requirements may demand the implementation of expert system technology to automate decision making. The analysis capabilities must be sufficiently robust not only to address sequencing of product but also to evaluate the need to reschedule based on comparison of actual and planned performance of the system.

Animation as a Formal Modeling Tool

Animation of industrial systems driven by simulation models has been available for some time (Pritsker, 1986). It has recently reached a high level of sophistication and availability to modelers because of the decreased hardware costs and increased capabilities of the hardware. High-resolution animation models can now easily be created with a variety of simulation software tools and integrated with the simulation model.

Significant research opportunities still exist in characterizing animation as a formal modeling tool. These opportunities exist primarily in the application of simulation for system design. Animation of a physical system is basically another way of modeling that system. When building a design model that will include animation, significantly more effort must be expended, for the capability needed to drive the animation is much greater than is needed to carry out analysis.

More research is needed to characterize the content of an animation relative to design objectives and also in interfacing that animation model efficiently with the analysis model. Research is also needed to minimize the added model development burden imposed by animation. A formal modeling procedure is needed for animation, including procedures for integration with the design or analysis model.

Additional research is also needed in defining the human interface requirements between the animation and the animation model builder. Most animation systems are somewhat limited in the amount of information they can display because of the size of the graphic screen available. Currently available features such as windows and divided screens help, but easier-to-use tools are required.

Animation is also a useful tool in describing the system's status in a real-time mode and triggering responses by the user to problems that exist in the system. It can be important in scheduling and control models to support the user in analyzing the dynamics of the proposed production schedule and identifying problems that might occur. However, production schedulers typically do not have the time required to review the variety of animations necessary to make effective use of this tool. Perhaps additional work to reduce the data displayed and the review time required might make animation an effective tool for production scheduling environments.

CONCLUSIONS

This paper discusses the effective use of simulation both for design applications by engineering groups and for scheduling and control applications by production managers. Simulation is a technology that is intuitively appealing to a wide audience and that can provide much insight in predicting the performance of industrial systems. Its growth in the future will depend on the effective use of both hardware and software technology and the integration of the technology with simulation software. New tools such as expert systems will help solve many of the problems. Cheaper and more powerful computer hardware will expand the availability of simulation technology to a wider group of users. Simulation remains a fertile area for development and provides many challenges for researchers.

REFERENCES

Grant, F. H. 1986. Production scheduling using simulation technology. Pp. 129–138 in Proceedings of the Second International Conference on Simulation and Manufacturing. Bedford, England: IFS Conferences, Ltd.

Grant, F. H. 1987. Scheduling and loading techniques. In Production and Inventory Control Handbook, 2d ed., F. H. Grant and J. H. Green, eds. Falls Church, Va.: American Production and Inventory Control Society.

Grant, J. W., and S. A. Weiner. 1986. Simulation Series, Part 4: Factors to consider in choosing a graphically animated simulation system. Industrial Engineering 18(8):37–38 and 65–68.

Haider, S. W., and J. Banks. 1986. Simulation Series, Part 3: Simulation software products for analyzing manufacturing systems. Industrial Engineering Errata 18(9):87.

Miner, R. J., and L. J. Rolston. 1983. MAP/1 User's Manual. West Lafayette, Ind.: Pritsker & Associates, Inc.

Musselman, K. J. 1984. Simulation: A design tool for FMS. Manufacturing Engineering 93(3):117–120.

Pritsker, A. A. B. 1986. Introduction to Simulation and SLAM II, 3d ed. New York: Halsted Press and West Lafayette, Ind.: Systems Publishing Corporation.

Pritsker, A. A. B., F. H. Grant, and S. D. Duket. 1986. Simulation in real-time factory control. Presented at a conference on Real-Time Factory Control, May 13–14, 1986. Dearborn, Mich.: Society of Manufacturing Engineers.

Reddy, Y. V., and M. S. Fox. 1982. KBS: An Artificial Intelligence Approach to Flexible Simulation. CMU-RI-TR-82-1. Robotics Institute, Carnegie Mellon University.

Standridge, C. R., and A. A. B. Pritsker. 1987. TESS: The Extended Simulation Support System. West Lafayette, Ind.: Pritsker & Associates, Inc.

THE HUMAN ROLE IN ADVANCED MANUFACTURING SYSTEMS

William B. Rouse

ABSTRACT   This paper is concerned with the conceptual design of support systems for humans in advanced manufacturing systems. A design methodology is presented that includes five major steps or phases: characterizing users' tasks, assessing demands of tasks, identifying approaches to support, determining likely obstacles, and anticipating user acceptance problems. This methodology is discussed in the context of its previous applications in aerospace and process control domains and its potential application in manufacturing.

INTRODUCTION

This paper considers a methodological framework based on an integrated view of human decision making in complex systems and explores its applicability to advanced manufacturing. This framework, which can be of great value for identifying potential problems and likely solutions in the design of advanced manufacturing systems, has been formalized and applied over the past 5 years (Rouse and Rouse, 1983; Rouse et al., 1984; Rouse, 1986). This framework has been used for almost 20 years in aerospace and in process and power systems, as well as for several applications in public service systems.

Many of the problems associated with the human role in complex systems are due to the technology-driven nature of most system development efforts. The "technology spiral" shown in Figure 1 illustrates the cen-tral phenomena in technology-driven developments (Rouse, 1985). An understanding of these phenomena helps to explain why the human role in complex systems becomes confusing.

The use of advanced technology is usually motivated by perceived performance or productivity requirements or the availability of new technology. Although the infusion of technology is the "standard" solution to most problems, the added technology may overwhelm the operators and managers who must work with it and those who must maintain it. This situation can be illustrated by the following examples. Pilots of F-15 fighter aircraft and B767 commercial aircraft can be overwhelmed by the number of modes of the radar system and the flight management system. In process and power systems, operators can be confused by the many failure modes of the automation that is supposed to help them. On

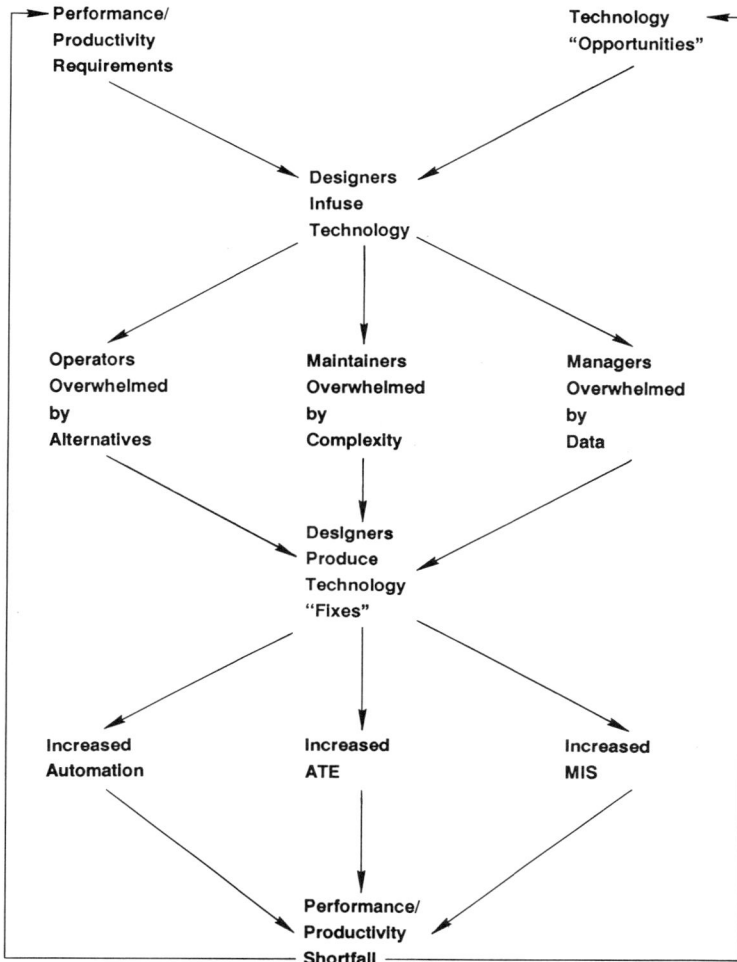

FIGURE 1   The technology spiral.

a high-tech injection molding control system, the introduction of color graphic displays and keyboards changed a relatively easy job into one that was quite difficult for the operators. Finally, of course, most of us have had frustrating experiences dealing with the multitude of cryptic commands for word processing and spreadsheet programs, with the typical result being that we learn a minimal set of commands and avoid the rest.

The plethora of functionality that can overwhelm an operator can also be a problem for those who must maintain the systems. More technology packed in smaller boxes in tighter spaces presents obvious difficulties in access and installation. Beyond these traditional maintainability problems, however, the complexity of advanced hardware and software requires that the troubleshooter have sophisticated knowledge and skills. This problem is exacerbated when complex subsystems are combined to produce integrated systems, such as for avionics or the multivariable control of large systems. The result is that increased maintenance hours are required per hour of operation, with a corresponding reduction in operational readiness or plant availability.

Although managers seldom directly operate and maintain complex systems, they are greatly affected by the infusion of technology into those systems. Computer and communications technologies allow collection, compilation, and transmission of enormous amounts of data. Much of this is not filtered and sampled in ways that provide managers with exactly the information they need. It is often unclear what information is needed to manage a complex system. As a result, everything imaginable and measurable is compiled, and managers tend to receive large amounts of data but little information.

The problems are evident in many domains; the issue is not whether they are real, but how to solve them. As shown in Figure 1, the technology-driven approach to system development dictates that these problems be "fixed" with more technology. The tendency is to eliminate operators by automation, robotize maintainers by automatic test equipment (ATE), and proceduralize managers by intelligent management information systems (MIS).

As might be expected, however, these types of technological fixes create the same types of problems as the technology infusions that produced the need for the fixes. As a result, performance and productivity shortfalls emerge, in part from technological problems and in part from inflated expectations. Although it might be imagined that this situation would cause the overall approach to be reconsidered, the more common response is to use the shortfall as a basis for reformulating requirements while also searching for new technological opportunities. Of course, this response ensures that the spiral continues!

Technology should be viewed as a means rather than an end in itself. From this perspective, the system development process should be objectives-driven, with technology considered only after objectives and requirements are formulated. Further, the *design* objectives should be oriented toward providing means to help users achieve the *operational* objectives for which they are responsible; i.e., they should be user-oriented.

Few people will disagree with this as a philosophical position. Everyone wants his or her system to be ergonomically designed, user-friendly, and so forth. However, many people believe that user-centered, objectives-driven design is not practical. To an extent, this point of view is due to a perception that one must first make sure that the technology works. Of course, by the time this assurance is received, a de facto commitment has usually been made to the technology of concern. Beyond this preoccupation with technology, there is also the feeling that the design tools and methods do not exist for achieving the well-intended objective of being "user-friendly." This criticism is well-founded since, until recently, human factors engineers, ergonomists, and engineering psychologists, among others, have had virtually no methodologies to contribute to the conceptual design phases of system development. The framework outlined in this paper is offered to ameliorate this deficiency.

BACKGROUND

A review of the current state of the art in advanced manufacturing systems and the salient issues being debated in the manufacturing community provides a context for some of the earlier statements while also motivating many of the methodological issues that are discussed later.

State of the Art

Twenty years ago, when an undergraduate student was provided the opportunity in a manufacturing course to program a numerically controlled (NC) milling machine, it was viewed as a high-tech adventure. Since then, NC evolved into computer-controlled NC (CNC), while computer-aided design (CAD) was evolving to include computer-aided manufacturing (CAM) and eventually CAD/CAM with an implicit direct link to CNC. Somewhere along the way, robots emerged to help with assembly and material handling, and this, when combined with CNC or CAD/CAM, led to flexible manufacturing systems (FMSs). Finally, combining all of these technologies with networked data bases for engineering, production, purchasing, and so forth, led to computer-integrated manufacturing (CIM) and the "lights-out factory of the future."

With these technological trends and the proliferation of acronyms came promises that manufacturing productivity and competitiveness would soar, leading to impressive "bottom lines" in the near future. As might be predicted from the earlier discussion, technology infusion was viewed as the panacea.

Although progress has been impressive, it certainly does not meet many expectations. It appears that CAD is well ensconced, integrated CAD/CAM is still in development, computer-aided engineering (CAE) is still in the laboratory, and CIM faces great challenges in implementation (Blumenthal and Dray, 1985). High downtime is a general problem (Shaiken, 1985), and inherent software incompatibilities across functional areas are impeding CIM (Conaway, 1985). "Islands of automation" are emerging (Lowndes, 1985), just when they might be expected to disappear, and plans for totally automated factories are running behind schedule (Lowndes, 1986). Nevertheless, solid technological progress is being made (Lowndes, 1985; Parks, 1987), and the technology spiral appears to be functioning.

Management Issues

Aside from the incremental progress in technology, advanced manufacturing technology presents many other issues. One of these is how this technology should be con-

sidered within the strategic plans of the company. The normal staff level approach to strategic planning appears to have problems with the technological discontinuity that can result when implementing the new technology. Much more than in the past, technology may have to be acquired from outside sources and adapted to particular applications, while also adopting a much longer term perspective on productivity improvements and return on investments (Ayres and Miller, 1983). It also appears that a "champion" is necessary if such relatively radical changes are to become part of the strategic plan (Meredith, 1986).

For advanced manufacturing technology to be effective, it may be necessary to rethink organizational structures. Traditional managerial and organizational contexts can impede innovation (Davis, 1986). Decisions concerning the degree of centralization versus decentralization can also be affected. In an analysis of the applicability of military command and control concepts to FMS organization, it was noted that decentralized structures tend to minimize the information processing requirements for each individual in the structure, while centralized structures can more easily recover from degraded operations (Armstrong and Mitchell, 1986).

A particularly interesting trend is the impact of advanced manufacturing technology on the role of middle management. Implementation of the new technology has tended to broaden the scope of jobs on the shop floor to include planning, diagnosing, operating, and maintenance (National Research Council, 1986) or, using somewhat different descriptors, analysis and programming (Fraser, 1986). As a result, operating decisions are being delegated to the working level (Fraser, 1986; National Research Council, 1986; Parks, 1987). With this additional responsibility and authority, as well as the requisite information, work teams are tending to displace middle management (National Research Council, 1986; Polakoff, 1987). It has been suggested that this dramatic reduction in the role of middle management in information compilation and management need not eliminate this level of personnel if their responsibilities can be shifted to emphasize communicating with employees, directing preventive maintenance, and improving quality control (Polakoff, 1987).

Until recently, applications of computer technology in manufacturing have been focused on reducing direct production costs. As a result, direct labor costs are becoming a decreasing percentage of total costs. The emphasis is therefore now shifting to using automation to cut overhead costs, including materials, which can account for 75 to 90 percent of the total cost (Dornheim, 1986; Lowndes, 1986). The implication of this trend is that the roles of both white-collar and blue-collar workers are being affected by advanced manufacturing technology.

Human Resources Issues

From the shop floor to upper management, humans should be viewed as resources within manufacturing systems, rather than potential Luddites. It is not simply a matter of being humane and socially minded. A lack of consideration of human resources issues can undermine the implementation of advanced manufacturing technology (Davis, 1986; Shaiken, 1985). Further, a recent study found that innovative human resources practices are often associated with what are judged to be successful implementation efforts (National Research Council, 1986). It has been argued that reluctance to embrace the technology could be assuaged by a general human resources policy that lessened uncertainty about job security by, for example, guaranteeing retraining (Ayres and Miller, 1983).

The issue of retraining is particularly problematic. There is widespread agreement that reeducation and retraining are essential to any human resources plan (American Management Association, 1986a, 1986b; Ayres and Miller, 1983; Margulies, 1985; Salvendy, 1985). It is argued that such training efforts are important for shop personnel, middle management, and all other levels including upper management (American Management Association, 1986a).

Although there appears to be a consensus on the need for training, the success of such efforts depends on there being an adequate population from which to select trainees. With the new technology, manufacturing jobs are changing to require more cognitive and reasoning abilities as well as levels of literacy that surpass mere reading and writing (American Management Association, 1986b). It has been estimated that less than 50 percent of current workers have the aptitudes to be trained for the new jobs (Salvendy, 1985). Considering new entrants into the manufacturing work force, the majority will be immigrants for whom at least the literacy requirements may present problems (American Management Association, 1986b). Ironically, at the same time that knowledge and skill requirements are outpacing the population of workers, many highly skilled machinists, albeit with different skills, are being relegated to the status of assembly-line workers for the purpose of monitoring CNC machines, robots, and FMS systems (Shaiken, 1985). These are important problems that require further examination and study (Committee on Science, Engineering, and Public Policy, 1987).

A final human resource issue concerns the resistance to change. It has been observed that the introduction of CIM systems has encountered stiff resistance (Blumenthal and Dray, 1985). Even well-intended, user-centered approaches to design and implementation have encountered substantial reluctance (Margulies, 1985). A useful insight is that change is a problem only in that it presents uncertainty (Nadler, 1986).

Human resource policies such as those discussed earlier may help to decrease uncertainty (Ayres and Miller, 1983). However, at the level of detail necessary for system design, much more concrete measures and methods are needed. An approach to anticipating and dealing with user acceptance problems is outlined in a later section of this paper.

System Design Issues

As noted in the introduction to this paper, there have been many proponents of user-centered system design in the manufacturing domain, but little methodology has been developed for pursuing this philosophy (Blumenthal and Dray, 1985; Brodner, 1985; Margulies, 1985). Nevertheless, various research efforts have provided insights and potential elements for more comprehensive methodologies.

Several related sets of guidelines have been developed that provide a taxonomy of the abilities and limitations of humans and computers, with particular emphasis on robotics and FMS applications (Hwang et al., 1984; Kamali et al., 1982; Nof et al., 1980). The primary use of these guidelines is in allocating functions between humans and computers. These types of guidelines were popular in aerospace applications in the 1950s and 1960s and a bit later in process control. It was found, however, that such taxonomies provide, at best, only a first step in resolving potentially difficult function allocation trade-offs. More recently, fairly sophisticated methods have been developed for function allocation (Rouse and Cody, 1986).

Design procedures are needed that provide a nominal sequence, but not a "cookbook," of issues to be resolved and decisions to be made that go beyond simple guidelines. Representative of a step in this direction in manufacturing is a series of model-based analyses of FMS scheduling. These efforts have included an analysis of the types of decision making in FMS scheduling (Fraser, 1986), the development of a hierarchical approach to integrating human capabilities into the decision-making process (Ammons, 1985), and an analytical decomposition of FMS cell schedule management and inventory management into six supervisory control subfunctions (Mitchell et al., 1986).

Methods and models should, to the extent possible, be evaluated relative to empirical data. Such data, unfortunately, are scarce for advanced manufacturing systems. Two studies of the use of computer-generated displays for FMS control concluded that the effects of the displays were much more subtle than anticipated (Mitchell and Miller, 1983; Sharit, 1984, 1985). Computer-generated information displays do not necessarily improve the performance of the operators and can degrade performance, unless additional assistance is provided for using the information in this form. Similar results have been found for procedure displays in flight management (Rouse and Rouse, 1980; Rouse et al., 1982) and searching aids in data-base retrieval (Morehead and Rouse, 1983). The allocation of responsibilities for machines between human and computer in FMS control has also been studied (Hwang, 1984).

Not only are modest amounts of data available, but they are also often plagued by considerable variability. Further, the trends in the data often defy intuition. It appears that this is a result of the nature of the problems being studied, as opposed to the skills of the investigators. It is simply not possible to study a system as complicated as FMS control by merely choosing naive experimental subjects and observing how they respond in a low-fidelity simulation. To be realistic and provide meaningful results, much more job training and aiding should be provided. In fact, the design and evaluation of such training and aiding are much more interesting and important issues than the effects of display formats, graphics, color coding, etc. (Govindaraj and Mitchell, 1985; Hwang, 1984; Mitchell and Miller, 1983).

Summary of Issues

The "factory of the future" is economically appealing although perhaps socially unsettling. The technological challenges rival, and may exceed, those in aerospace and process systems. Because of the complexity and costs involved, humans will still have many, if not more, roles in manufacturing systems. Fewer people may be doing more, which would seem to portend increased productivity. However, the level of understanding of the role of the human in manufacturing, as well as the tools and methods for supporting those roles, is severely limited. The remainder of this paper outlines an approach for helping to overcome this deficiency.

A FRAMEWORK FOR USER-CENTERED DESIGN

The important motivations for the user-centered design point of view fit in two classes: (1) human abilities and (2) human inclinations. Human manipulative skills continue to surpass those of automation, particularly in adaptability and flexibility, as contrasted with precision and consistency. Considering the ways in which advanced manufacturing technologies are affecting jobs, the perceptual, judgmental, and creative abilities of humans are likely to be more important than manipulative skills as reasons for humans to retain central roles in manufacturing.

Various theorists and practitioners in artificial intelligence continue to assert that computers will eventually supplant almost all human manipulative, perceptual, judgmental, and creative abilities. Even if this is true, which I strongly doubt, the inclinations of humans are such that they will continue to have important roles in com-

plex systems. In particular, the inclination of humans to make commitments and accept responsibility for their actions, as well as the actions and well-being of others, make them unique relative to imaginable hardware and software alternatives. Thus, I believe that user-centered design must not be considered a "holding action" until total automation is possible.

There are four key conceptual elements of the framework for user-centered design. The initial element relates to understanding user-system tasks and characterizing these tasks by means of a constrained set of terminology. This terminology provides direct links to the next two conceptual elements, which relate to identifying means for enhancing human abilities and overcoming human limitations for the tasks of interest. The fourth and final element relates to fostering user acceptance of the means identified for enhancing abilities and overcoming limitations. These four conceptual elements are the basis for the design methodology summarized in Figure 2 and described in the remainder of this section.

Characterizing Users' Tasks

Traditional approaches to describing users' tasks include direct observations and interviews for tasks where job incumbents are available and analytical decomposition of task scenarios for those that are new and not directly analogous to similar existing tasks. The value of these methods is limited during the early stages of design when a system concept is not yet available. There is a need for general procedures by which designers can choose a subset of tasks that appear to be the most important for the application domain of concern. By proceeding in this way, the roles for the humans become the first issue studied rather than the last.

With a goal of developing a general set of tasks, 120 publications that reported on decision-making and decision-support sys-

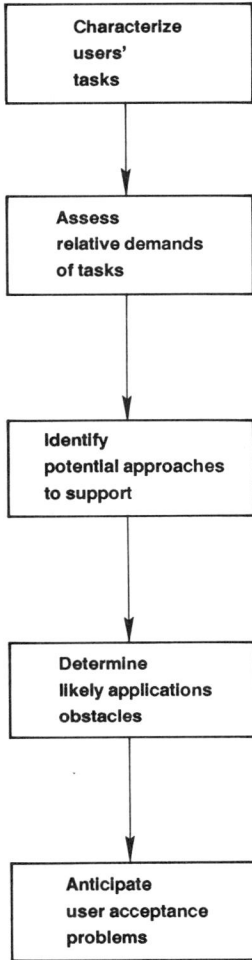

FIGURE 2   User-centered design methodology.

the solution will be acceptably close to the desired results.

As one might expect, Figure 3 is much too simple, particularly in relation to the interactions among tasks. Figure 4 is a refinement and expansion of the simpler representation. First, it reflects the fact that the a priori situation and human stereotypical plans and expectations govern much of behavior. Most situations and subsequent behaviors are fairly routine and, fortunately, considering the effort involved, situation assessment as well as planning and commitment need not be invoked. Occasionally, the consequences of actions deviate significantly from expectations, and routine behaviors are insufficient. Humans then must be concerned with explaining the observed deviations and choosing among alternative courses of action, whereupon the situation may, or may not, revert to routine.

Another important distinction shown in Figure 4 is the differentiation of task behaviors from the user-computer interface. User-centered design is much more than the analysis and synthesis associated with the

tems, primarily in the aerospace industry, were reviewed (Rouse and Rouse, 1983). All of these studies were concerned with one or more of the general tasks shown in Figure 3. These can be summarized as follows: Situation assessment is concerned with the formulation of the problem or deciding what is happening; planning and commitment are concerned with devising a solution to the problem or deciding what to do about it; and execution and monitoring involve implementing the solution or plan and determining whether the consequences of

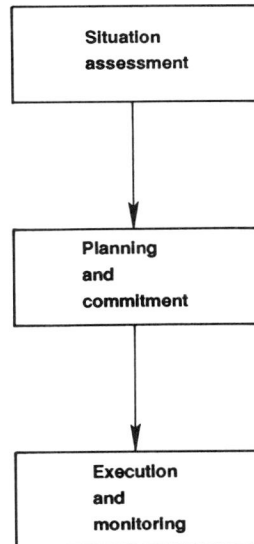

FIGURE 3   Three general user-system tasks.

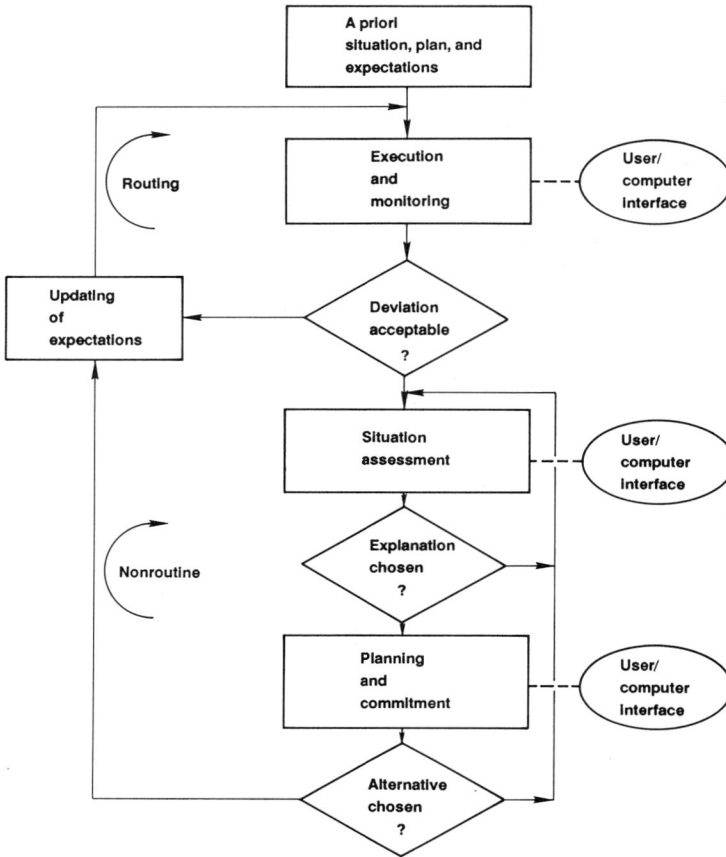

FIGURE 4   Relationships among user-system tasks.

hardware and software for displays and controls. Although these aspects of design are important during the later, more detailed stages of design, they will not be addressed in this paper, which is primarily concerned with the support of a structured approach to problem formulation and requirements analysis.

The process shown in Figure 4 can be further decomposed into the set of 13 tasks in Figure 5. This set of tasks was sufficient to classify and describe all decision-making and decision-support efforts reviewed (Rouse and Rouse, 1983). It is reasonable to ask, however, whether this set of tasks is unique to the aerospace applications from which it emerged.

In an effort to test the applicability of this task taxonomy to other problems, it was used to describe a set of decision-support systems in the process control domain (Rouse et al., 1984). Two independent analysts reviewed the documentation for several existing support systems and classified the functionality of these systems using Figure 5. The results of these two independent analyses for the process control example were virtually identical. Although the applicability of this task taxonomy to both aerospace systems and process control does not ensure its applicability to discrete parts manufacturing, my perception is that this taxonomy is applicable to advanced manufacturing systems. The accuracy of this perception can be finally demonstrated only through repeated testing of this approach in a variety of manufacturing circumstances.

Two characteristics of Figure 5 are of particular importance. First, most of the tasks involve generation, evaluation, and selection among alternatives. The use of this standard terminology will be shown to be helpful for identifying approaches to enhancing abilities and overcoming limitations in these tasks. The second noteworthy characteristic of Figure 5 is the emphasis on alternative interpretations of deviations, information sources, explanations, and courses of action. Thus, the structure of Figure 5 is

based on both the process depicted in Figure 4 and a three-by-four array of action words and objects of action. These complementary methods of organization bring an important degree of structure to the process of characterizing the tasks of the users.

Assessing Relative Demands of Tasks

It could easily be argued that every task involves all of the elements shown in Figure 5. Although this may be true, such a conclusion does not help in identifying the "bottlenecks" in user-system performance. Table 1 was prepared as an aid for determining which tasks are likely to present problems. This table is meant to prompt the thinking of the analyst rather than replace it. Context-specific knowledge should, of course, preempt any of the assessments in this chart.

To illustrate the use of Table 1, it will be useful to consider the emerging roles of the FMS operator. In his or her original job as a machinist, execution and monitoring were dominant, with occasional elements of the other tasks noted as "moderate" in the operations column of Table 1. More recently, CNC, CAD/CAM, and FMS have caused the machinist's job to include more of a maintenance role, involving aligning parts, clearing debris, and dealing with snags in the flow of parts and materials (Shaiken, 1985). From Table 1, one can see that this lessens the relative demands for some of the more cognitive tasks, thereby increasing the requirements for manipulative and perceptual skills as contrasted with judgmental and creative abilities.

It would appear, however, that the eventual role of the FMS operator will increase the demands on judgmental and creative abilities, in particular by shifting the emphasis to elements of planning and commitment (Ammons, 1985; Fraser, 1986; Mitchell et al., 1986). Thus, the emerging roles of the FMS operator would appear to include some aspects of the roles of manage-

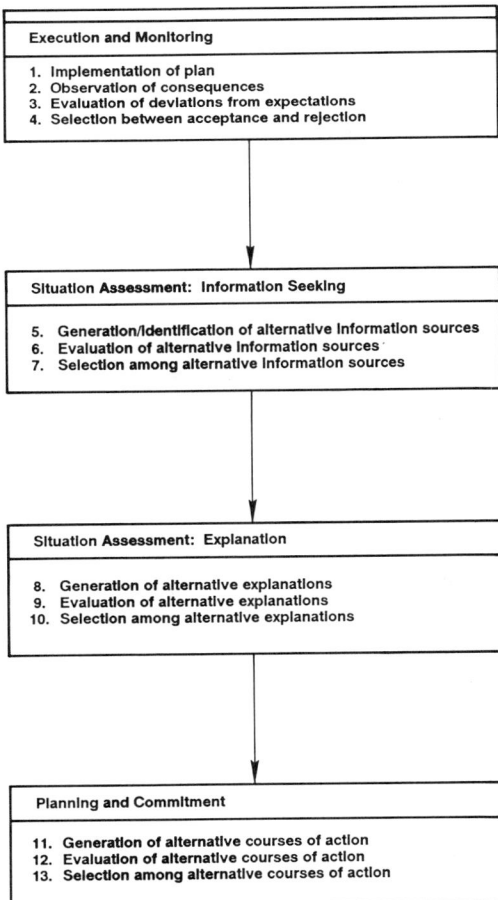

FIGURE 5 Subtasks of general user-system tasks.

TABLE 1 Relative Demands of User-System Tasks

User-System Tasks	Type of User			
	Operations	Maintenance	Management	Design
Execution and monitoring				
Implementation	High	High	Low	Low
Observation	High	Moderate	Low	Low
Evaluation	Moderate	Moderate	Moderate	Moderate
Selection	Moderate	High	High	Moderate
Situation assessment:				
Information seeking				
Generation	Low	Low	High	Moderate
Evaluation	Moderate	Low	High	High
Selection	Moderate	Moderate	Moderate	High
Situation assessment:				
Explanation				
Generation	Moderate	Moderate	High	Low
Evaluation	Low	Moderate	High	Low
Selection	Low	Moderate	Moderate	Low
Planning and commitment				
Generation	Low	Low	High	High
Evaluation	Moderate	Moderate	High	High
Selection	Moderate	Moderate	Moderate	Moderate

ment and design. My perception is, however, that the projected demands for situation assessment indicated for the management and design roles shown in Table 1 will be "moderate" rather than "high" for the FMS operator. This illustrates the earlier point that context-specific knowledge should preempt the entries in Table 1.

This brief analysis suggests that technology will initially move the work content for the machinist toward that of lower-level positions, comparable to that of assembly workers (Shaiken, 1985), but that these positions will eventually evolve to a higher level, such as FMS cell supervisors (Ammons, 1985; Fraser, 1986; Mitchell et al., 1986). This trend will obviously require great flexibility from the workers. More central to the theme of this paper, however, is the conclusion that the changing roles of these workers dictate changing approaches in assisting them to achieve their operational objectives.

Identifying Approaches to Support

At this point in an analysis, Figure 5 and Table 1 have been used in conjunction with domain-specific knowledge to identify one or more tasks that appear most in need of support. The design team could now sit around a table and brainstorm to produce support concepts. However, such an approach would ignore the thousands of previous efforts to develop support systems. What is needed is an easy method of accessing these previous efforts.

Figures 6 and 7 were synthesized from ongoing reviews of hundreds of support system development and evaluation projects as well as many years of experience in support systems R&D. Although these tabulations are useful for prompting ideas, they are also used to access a card file and, subsequently, a small library of documents on support systems. This method of identifying and retrieving information could easily be com-

GENERATION OF ALTERNATIVES
o   For a given situation, a support system might retrieve previously relevant and useful alternatives.
o   For a given set of attributes, a support system might retrieve candidate alternatives with these attributes.
o   Given feedback with regard to suggested alternatives, a support system might adapt its search strategy and/or tactics.

EVALUATION OF ALTERNATIVES
o   For a given alternative, a support system might assess the alternative's a priori characteristics such as relevance, information content, and resource requirements.
o   For a given situation and alternative, a support system might assess the degree of correspondence between situation and alternative.
o   For a given alternative, a support system might assess the likely future consequences such as expected impact and resource requirements.
o   For given multiple alternatives, a support system might assess the relative merits of each alternative.
o   Given feedback of appropriate variables, a support system might adapt its evaluations in terms of time horizon, accuracy, etc.

SELECTION AMONG ALTERNATIVES
o   For given criteria and set of evaluated alternatives, a support system might suggest the selection that yields the "best" allocation of human and system resources.
o   For given individual differences and time-variations of criteria, preferences, and evaluations, a support system might adapt its suggestions to reflect these variations.

FIGURE 6   Alternative approaches to supporting generation, evaluation, and selection of user systems

puterized, but the investment in such an effort cannot be justified until there are more users of this information.

Many of the entries in Figures 6 and 7 are self-explanatory, but a few clarifications are needed. Although the entries concerned with the generation of alternatives appear straightforward, the suggested support is difficult to provide (Madni et al., 1985) in that there are few previous efforts to draw upon. This difficulty appears to be due, for the most part, to the difficulty in specifying the attributes that are desired. Some progress has been made in using pattern-recognition methods to infer attributes of desired alternatives from a set of examples (Freedy et al., 1985; Morehead and Rouse, 1985). It is fairly straightforward to retrieve the examples if users can define them appropriately—e.g., by requesting information on all flight directors for jet fighter aircraft or all of the types of robot manipulators that are currently available.

The evaluation of alternatives is easy to understand in that this type of activity is common in engineering analysis. The feasibility of supporting evaluation depends on the availability of appropriate models, cal-

INPUTS TO THE USER
o   For given information, a support system might transform, format, and code the information to enhance human abilities and overcome human limitations.
o   For a given set of evaluated information, a support system might filter and/or highlight the information to emphasize the most salient aspects of the information.
o   For a given sample of information, a support system might fit models to the information in order to integrate and interpolate within the sample.
o   For given constraints and individual differences, a support system might adapt transformations, models, etc.

OUTPUTS FROM THE USER
o   For a given plan and information regarding the user's actions, a support system might monitor implementation for inconsistencies and errors of omission and commission.
o   For a given plan and information regarding the user's actions and intentions, a support system might perform some or all of the implementation to compensate for the user's inconsistencies, errors, or lack of resources.
o   Given information on intentions, resources available, priorities, etc., a support system might adapt its monitoring and/or implementation.

FIGURE 7   Alternative approaches to supporting the input to users and the output from the users of systems.

culation techniques for the alternatives, and measures of interest. Finite difference methods and geometric modeling techniques in CAD/CAM/CAE represent evaluation supports for designers. Spreadsheet models provide similar support for managers.

The majority of previous efforts to develop support systems have focused on the selection among alternatives, in part because this type of support is most tractable. If all of the alternatives have been specified and the probability distributions associated with the consequences of choosing each alternative are known and the decision makers' criteria can be assessed, it is usually easy to determine the best or optimal alternative. For alternatives involving multiple stages, locations, or the like, this optimization problem is less straightforward but can, nevertheless, be treated using standard control theory and operations research techniques. Although the techniques used to

support selection among alternatives are important, experience suggests that identification of feasible alternatives and their likely consequences is often sufficient for decision makers to choose immediately without resorting to optimization. Thus, despite the great attention that has been given to selection among alternatives, this task is usually not the most difficult task faced by humans. Good support for generation and evaluation is typically more important but unfortunately is seldom available.

Although the support of generation, evaluation, and selection is central to this user-centered design framework, these types of support are not sufficient for a comprehensive approach to user-centered design. Figure 7 provides guidance in choosing approaches for supporting inputs to the user and outputs from the user. With regard to inputs, display design has long been the stock in trade of human factors engineers.

Fairly recently, methods based on expert systems technology have been developed for on-line, intelligent information management. These methods have the potential to filter, transform, and format information automatically. This capability will, for example, make it feasible for operators of complex systems to cope with the enormous amounts of data available through computer and communications technologies (Rouse et al., 1986). On the output side, it is feasible to have a computer monitor action sequences for reasonableness and consistency. An example of this support concept is an onboard information system for aircraft that was developed and evaluated a few years ago (Rouse et al., 1982). More recently, this concept has been generalized to a comprehensive architecture for error-tolerant interfaces (Rouse and Morris, 1987).

To summarize this discussion of identifying approaches to supporting users, all of the vast support system literature describes theories, design concepts, and evaluative results that relate to one or more of the capabilities summarized in Figures 6 and 7. Support concepts for selection and input are the most common; concepts for evaluation and output are not uncommon; and concepts for generation are fairly rare.

Determining Likely Obstacles

Although there is a wealth of support concepts, there are several types of obstacles that can limit feasibility of applying these ideas. Table 2 provides some guidance for identifying likely obstacles. Potential problems with the nature of the knowledge base relate to the extent to which it is likely to be difficult to understand and perhaps model or capture knowledge and skills for each type of user. Obstacles associated with the nature of the interaction between the user and the system are related to the difficulties of developing flexible and fast intelligent interfaces. There are also risks associated with inadequate resolution of the knowledge base and interaction problems. The consequences are usually immediate and can be traced for operations and maintenance. For management and design they are typically delayed and are not traceable.

Although Table 2 is not comprehensive and does not provide guidance for overcoming major obstacles, this compilation of experiences is useful for determining the scope required for a support system development effort. In other words, Table 2 can be used as a guide for budgeting resources in anticipation of likely obstacles. As noted before, this type of chart should be used to

TABLE 2 Obstacles to Application of Support Concepts

Potential Problems	Type of User			
	Operations	Maintenance	Management	Design
Nature of knowledge base				
Lack of structure	Low	Low	High	Moderate
Less than comprehensive	Low	Moderate	High	High
Inaccessible	Moderate	Moderate	High	Moderate
Nature of interaction				
Heterogeneity of users	Low	Moderate	High	Moderate
Unacceptability of prescriptions	Moderate	Low	High	High
Real-time requirements	High	Moderate	Low	Low
Nature of risks				
Potential immediate misfortune	High	Moderate	Low	Low
Potential long-term misfortune	Low	Moderate	High	High
Lack of traceability	Low	Low	High	Moderate

prompt questions rather than as a replacement for thinking about the specifics of the application of interest.

Anticipating User Acceptance Problems

The review of R&D efforts in support systems suggested that few support system concepts are ever used operationally. Studies are performed, reports are written, and the effort is frequently terminated. In attempting to understand this situation further, it was found that many of the support systems that are fielded encounter indifference or opposition from the personnel for whom the support is intended. This finding led us to a detailed examination of user acceptance (Rouse and Morris, 1986).

There appear to be four determinants of user acceptance. Most obvious is the users' perceptions of the impact of the support system on the quality of their job performance. In other words, does the system work as advertised and, if so, does this functionality help the individual user? Another fairly obvious dimension is the perceived ease of use. Specifically, is the perceived effort required to learn and use the support system outweighed by the perceived benefits?

The remaining dimensions are more subtle. One of these is the perceived impact on desired levels of discretion. In particular, is the support system such that users retain the desired opportunities to exercise their manipulative, perceptual, judgmental, and creative skills? The final dimension is perceived peer group and organizational attitudes toward the support system. Users are more likely to accept a support system if their colleagues and their supervisors extol its features and benefits.

The development of this four-dimensional view of user acceptance led to the development of a structured approach for anticipating user acceptance problems and attempting to resolve them in parallel with the overall design process. This structured

approach is shown in Figure 8. The process starts with a set of candidate functions in which automation or other technology infusions appear feasible and warranted. The process then proceeds through the structured set of considerations shown in Figure 9 to prune and modify the candidate functions, while also preparing and involving eventual users in planning and implementing the changes. The basis for the entries in Figure 9 is discussed in detail elsewhere (Rouse and Morris, 1986). Within the scope of this paper, it is sufficient to indicate the availability of a method for dealing with the important problem of user acceptance.

Summary of Framework

The user-centered design framework presented in this section is basically a structured approach for characterizing users' tasks, assessing relative demands, identifying approaches to support, determining

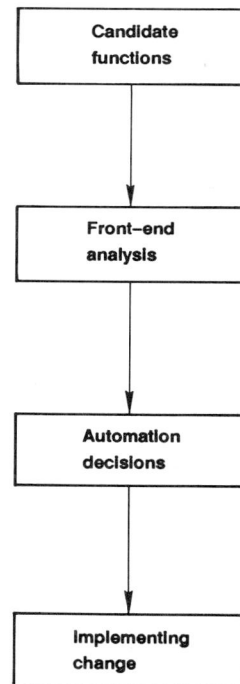

FIGURE 8   Fostering user acceptance.

FRONT-END ANALYSIS

1. Characterize the functions of interest in terms of whether or not these functions currently require humans to exercise significant levels of skill, judgment, and/or creativity.

2. Determine the extent to which the humans involved with these functions value these opportunities to exercise skill, judgment, and/or creativity.

3. Determine if these desires are due to needs to feel in control, achieve self-satisfaction in task performance, or perceptions of potential inadequacies of automation technology in terms of quality of performance and/or ease of use.

4. If need to be in control or self-satisfaction are not the central concerns, determine if the perceived inadequacies of the automation technology are well founded; if so, eliminate the functions in question from the candidate set -- if not, provide demonstrations or other information to familiarize personnel with the actual capabilities of the automation technology.

AUTOMATION DECISIONS

5. To the extent possible, only automate the system functions that personnel in the system feel should be computerized or computer aided (i.e., those for which they are willing to lose discretion).

6. To the extent necessary, particularly if number 5 cannot be followed, consider increasing the level and number of functions for which personnel are responsible so that they will be willing to delegate the functions of concern (i.e., expand the scope of their discretion).

7. Assure that the level and number of functions allocated to each person or type of personnel form a coherent set of responsibilities, with an overall level of discretion consistent with the abilities and inclinations of the personnel.

8. Avoid automating functions when the anticipated level of performance is likely to result in regular intervention on the part of the personnel involved (i.e., assure that discretion once delegated need not be reassumed).

IMPLEMENTING CHANGE

9. Assure that all personnel involved are aware of the automation effort and what their roles will be after the change.

10. Provide training that assists personnel in gaining any newly required abilities to exercise skill, judgment, and/or creativity and helps them to internalize the personal value of having these abilities.

11. Involve personnel in planning and implementing the changes from both a system-wide and individual perspective, with particular emphasis on making the implementation process minimally disruptive.

12. Assure that personnel understand both the abilities and limitations of the increased automation, know how to monitor and intervene appropriately, and retain clear feelings of still being responsible for system operations.

FIGURE 9   An approach for anticipating and avoiding user acceptance problems.

likely obstacles, and anticipating user acceptance problems. Various subsets of these five components of this methodology have been applied to several applications in the aerospace and process control domains. They are now being used for several new applications in these domains.

The framework is basically a user-centered approach to problem formulation and requirements analysis. Once the process outlined in this paper is completed, there is still a tremendous amount of design work left to be done. This work includes development of the functional architectures of the support system concepts chosen, integration of these architectures into the overall manufacturing system architecture, and completion of the detailed design. Methodological support for the more detailed aspects of the user-system design has been developed and is compatible with the methods presented here. The outputs of the process described in this paper serve as inputs to more detailed design procedures (Rouse et al., 1984; Rouse, 1986).

CONCLUSIONS

Manufacturing is undergoing a metamorphosis, more slowly than anticipated, but nevertheless it is progressing. Associated with this substantial change are three important trends. First, information technology is replacing physical technology as the central concern. Second, as a result of the first trend, software is replacing hardware as the key to productivity. Third, cognition and reasoning abilities are replacing sensorimotor skills as the raison d'être for the human role in manufacturing systems. Similar trends are evident and more mature in the aerospace domain and, to a lesser extent, the process and power industries. Thus, the management, human resources, and design issues that these trends pose for the manufacturing industry are neither novel nor unique. The industry, therefore, can benefit from some well-reasoned technology transfer.

This technology should not be limited to hardware and software. There is a strong need for methodology in general, and user-centered design methodology in particular. Clearly, traditional methods of ergonomics and safety engineering are no longer sufficient, and often not even appropriate, for

understanding and supporting the human role in advanced manufacturing systems. This paper has suggested a new way of looking at user-system problems, as well as a structured approach for resolving these problems. Although tailoring may be necessary to fit the manufacturing context, it appears that transferring this technology to manufacturing would be a good first step.

REFERENCES

American Management Association. 1986a. Report of the manufacturing council. AMA Council Reports (Winter):5–6.

American Management Association. 1986b. Report of the manufacturing council. AMA Council Reports (Summer):9.

Ammons, J. C. 1985. Scheduling models for aiding real time FMS control. Pp. 185–189 in Proceedings of the 1985 IEEE International Conference on Systems, Man, and Cybernetics.

Armstrong, J. E., and C. M. Mitchell. 1986. Organizational performance in supervisory control of flexible manufacturing systems. Pp. 1437–1442 in Proceedings of the 1986 IEEE International Conference on Systems, Man, and Cybernetics.

Ayres, R. U., and S. M. Miller, eds. 1983. Robotics: Applications and Social Implications. Cambridge, Mass.: Ballinger.

Blumenthal, M., and J. Dray. 1985. The automated factory: Vision and reality. Technology Review (January):28–37.

Brodner, P. 1985. Qualification based production: The superior choice to the "unmanned factory." Pp. 18–22 in Proceedings of the 1985 IFAC Conference on Analysis, Design, and Evaluation of Man-Machine Systems.

Committee on Science, Engineering, and Public Policy. 1987. Technology and Employment: Innovation and Growth in the U.S. Economy, R. M. Cyert and D. C. Mowery, eds. Washington, D.C.: National Academy Press.

Conaway, J. 1985. Integrated data flow in CIM systems. IE News on Computer and Information Systems 20:1–4.

Davis, D. D., ed. 1986. Managing Technological Innovation. San Francisco: Jossey-Bass.

Dornheim, M. A. 1986. Airframe makers expect computer techniques to cut overhead costs. Aviation Week & Space Technology (December 22):56–59.

Fraser, J. M. 1986. Effects of flexible, computerized manufacturing systems on decision making. Pp. 1303–1306 in Proceedings of the 1986 IEEE International Conference on Systems, Man, and Cybernetics.

Freedy, A., A. Madni, and M. Samet. 1985. Adaptive user models: Methodology and application in man-computer control. Pp. 249–293 in Advances in Man-Machine Systems Research: 2, W. B. Rouse, ed. Greenwich, Conn.: JAI Press.

Govindaraj, T., and C. M. Mitchell. 1985. Decision support systems for real time control of flexible manufacturing systems. Pp. 56–58 in Proceedings of the 1986 IEEE International Conference on Systems, Man, and Cybernetics.

Hwang, S. 1984. Human supervisory performance in flexible manufacturing systems. Doctoral dissertation, Purdue University.

Hwang, S., W. Barfield, T. Chang, and G. Salvendy. 1984. Integration of humans and computers in the operations and control of flexible manufacturing systems. International Journal of Production Research 22:841–856.

Kamali, J., C. L. Moodie, and G. Salvendy. 1982. A framework for integrated assembly: Humans, automation, and robots. International Journal of Production Research 20:431–448.

Lowndes, J. C. 1985. Lockheed installs advanced facilities as part of factory modernization. Aviation Week & Space Technology (May 27):113–118.

Lowndes, J. C. 1986. Management by computer promises unprecedented productivity gains. Aviation Week & Space Technology (December 22): 50–55.

Madni, A., M. Brenner, I. Costea, D. MacGregor, and F. Meshkinpour. 1985. Option generation: Problems, principles, and computer-based aiding. Pp. 757–760 in Proceedings of the 1985 IEEE International Conference on Systems, Man, and Cybernetics.

Margulies, F. 1985. Flexible automation: New options for men, economy, and society. Pp. 14–17 in Proceedings of the 1985 IFAC Conference on Analysis, Design, and Evaluation of Man-Machine Systems.

Meredith, J. R. 1986. Strategic planning for factory automation by the championing process. IEEE Transactions on Engineering Management EM-33:229–232.

Mitchell, C. M., and R. A. Miller. 1983. Design strategies for computer-based information displays in real-time control systems. Human Factors 25:353–369.

Mitchell, C. M., T. Govindaraj, O. Dunkler, S. P. Krosner, and J. C. Ammons. 1986. Real time scheduling in FMS: A supervisory control model of cell operator function. Pp. 1443–1448 in Proceedings of the 1986 IEEE International Conference on Systems, Man, and Cybernetics.

Morehead, D. R., and W. B. Rouse. 1983. Human-computer interaction in information seeking tasks.

Information Processing and Management 19:243–253.

Morehead, D. R., and W. B. Rouse. 1985. Computer-aided searching of bibliographic data bases: Online estimation of the value of information. Information Processing and Management 21:387–399.

Nadler, G. 1986. Breakthrough thinking for the integration engineer. Industrial Engineering (December:)22–25.

National Research Council (NRC). 1986. Human Resource Practices for Implementing Advanced Manufacturing Technology. Washington, D.C.: National Academy Press.

Nof, S. Y., J. L. Knight, and G. Salvendy. 1980. Effective utilization of industrial robots: A job and skills analysis approach. AIIE Transactions 12:216–225.

Parks, M. W. 1987. Expert systems: Filling the missing link in paperless aircraft assembly. Industrial Engineering (January):37–45.

Polakoff, J. C. 1987. Will middle managers work in the "factory of the future?" Management Review (January):50–51.

Rouse, S. H., and W. B. Rouse. 1980. Computer-based manuals for procedural information. IEEE Transactions on Systems, Man, and Cybernetics. SMC-10:506–510.

Rouse, S. H., W. B. Rouse, and J. M. Hammer. 1982. Design and evaluation of an onboard computer-based information system for aircraft. IEEE Transactions Systems, Man, and Cybernetics. SMC-12:451–463.

Rouse, W. B. 1985. The role of human factors in military R&D. Pp. 167–178 in Using Psychological Science: Making the Public Case, F. Farley, and C. H. Null, eds. Washington, D.C.: Federation of Behavioral, Psychological, and Cognitive Sciences.

Rouse, W. B. 1986. Design and evaluation of computer-based decision support systems. Pp. 259–284 in Microcomputer Decision Support Systems: Design, Implementation, and Evaluation, S. J. Andriole, ed. Wellesley, Mass.: QED Information Sciences.

Rouse, W. B., and W. J. Cody. 1986. Function allocation in manned systems. Pp. 1600–1606 in Proceedings of the 1986 IEEE International Conference on Systems, Man, and Cybernetics.

Rouse, W. B., and N. M. Morris. 1986. Understanding and enhancing user acceptance of computer technology. IEEE Transactions on Systems, Man, and Cybernetics SMC-16:965–973.

Rouse, W. B., and N. M. Morris. 1987. Conceptual design of a human error tolerant interface for complex engineering systems. Automatica 23:231–235.

Rouse, W. B., and S. H. Rouse. 1983. A framework for research on adaptive decision aids. Technical Report AFAMRL-TR-83-082. Wright-Patterson Air

Force Base, Ohio: Air Force Aerospace Medical Research Laboratory.

Rouse, W. B., R. A. Kisner, P. R. Frey, and S. H. Rouse. 1984. A method for analytical evaluation of computer decision aids. Technical Report NUREG/CR-3655; ORNL/TM-9068. Oak Ridge, Tenn.: Oak Ridge National Laboratory.

Rouse, W. B., N. D. Geddes, and R. E. Curry. 1986. An architecture for intelligent interfaces: Outline of an approach to supporting operators of complex systems. Pp. 914–920 in Proceedings of the 1986 National Aerospace Electronics Conference.

Salvendy, G. 1985. Human factors in planning robotic systems. Chapter 32 in Handbook of Industrial Robotics, S. Y. Nof, ed. New York: Wiley.

Shaiken, H. 1985. The automated factory: The view from the shop floor. Technology Review (January):17–24.

Sharit, J. 1984. Supervisory control of a flexible manufacturing system: An exploratory investigation. Ph.D. dissertation, Purdue University.

Sharit, J. 1985. Supervisory control of a flexible manufacturing system. Human Factors 27:47–59.

MODELING IN THE DESIGN PROCESS

Herbert B. Voelcker

ABSTRACT   This paper summarizes the evolution of modeling technology and provides a status report on the newest technology—solid modeling. It discusses two (of many) issues that indicate how little we really know about design and about the interplay between design and manufacturing, and it closes with the following assessment: Contemporary modeling systems are most useful for refining and documenting nearly finished designs and for driving a growing array of computer-aided manufacturing modules; they provide little help in the early, conceptual stages of design. Thus, we have, in essence, a growing technological imbalance, with manufacturing striding ahead of design in terms of scientific understanding and automation. The appendix explores the current situation by tracing some of the history of design and manufacturing.

INTRODUCTION

Design is the first major step in a product's life cycle, and design is often the main determinant of a product's manufacturability, salability, serviceability, and longevity. The quality of design in individual companies, specific industries, and whole nations is influenced by many factors, with the tools that are available to designers being among the most important. This paper focuses on modeling tools for discrete-goods design (loosely defined as mechanical design). These tools have evolved rapidly, but they still have major deficiencies, and design itself is poorly understood in a scientific sense.

Figure 1 shows an idealized product cycle. Sales and marketing define a new or revised product in terms of functional requirements, price or volume trade-offs, and other similar parameters. Design and engineering convert the set of perceived needs

and market constraints into complete specifications—a design—for a deliverable product. Manufacturing planners then produce specifications for the product's manufacture (typically process and inspection plans, numerical control [NC] programs, and the like), and these are executed to produce a product that is then marketed.

Design is often the pivotal operation in the product cycle because it establishes a match (a compromise) between the initial marketing goals and a product's "deliverable functionality," economic producibility, maintainability, and longevity. Clearly, the design capabilities of individual companies and whole nations are strong determinants of their long-term viability in a competitive world.

Although the companies' and nations' design capabilities are influenced by a host of commercial, cultural, and historical factors, the primary intrinsic determinants are

167

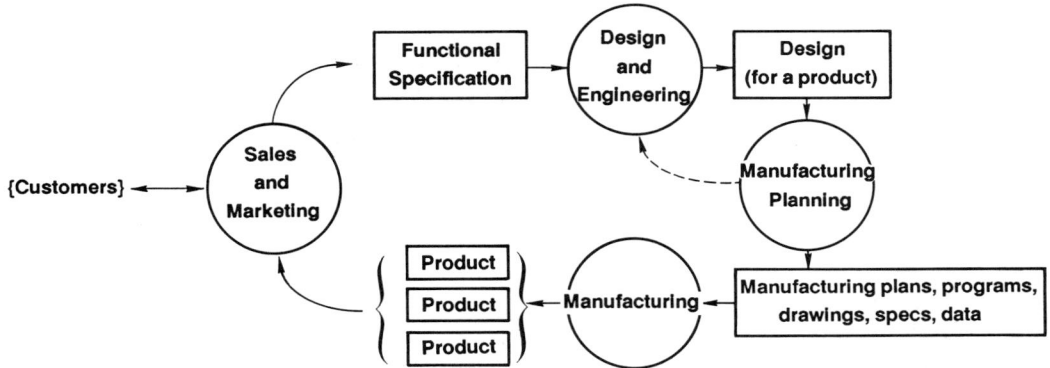

FIGURE 1   An idealized product cycle.

the skills of the designers and the tools and methods that they use. The principal focus in this paper is on tools—specifically, modeling tools—because these are understood well enough to admit technical assessment and forecasting. Although design methods and designers' skills are at least as important as tools, they are poorly understood and are covered only briefly and somewhat obliquely.

Further, the paper focuses on modeling tools for discrete-goods design (loosely, mechanical design—the term used hereafter for brevity) because mechanical design is both pervasive and ill-understood in a scientific sense. Nonmechanical design domains, such as digital electronics, chemical processes, and soft goods (e.g., apparel), pose unique and interesting problems, but each is smaller than the mechanical aggregate, and at least the first two are more advanced than mechanical design in terms of scientific understanding and automation.

Computer-aided design (CAD) and computer-aided manufacturing (CAM) systems have proliferated in the mechanical industries over the past two decades, and within each lies a modeling system of some kind. Although the early progress in CAD and CAM was paced mainly by advances in computing and graphics technology, progress in the past decade has been paced

mainly by advances in modeling and in understanding of how to use models.

This paper surveys the evolution and current status of mechanically oriented computer modeling and discusses two (of many) issues that indicate how little we know about design and about the interplay between design and manufacturing. It concludes that modern CAD/CAM systems are best suited to the final "tuning" and "detailing" of parts and products and as sources of data for increasingly automated manufacturing processes; they provide little help in the early, conceptual phases of design. The appendix traces some of the history of design and manufacturing in an effort to understand why manufacturing is ahead of design in terms of scientific understanding and automation.

MECHANICALLY ORIENTED MODELING SYSTEMS[1]

Contemporary modeling systems are concerned primarily with geometry. They

---

[1]The section dealing with mechanically oriented modeling systems is based in part on updated material from two papers written by A. A. G. Requicha and the author about 5 years ago (Requicha and Voelcker, 1982, 1983). These papers provide more than 150 references, most of which are still pertinent. Wolfe et al. (1987) provide a view from within the context of a single modeling system, namely, IBM's internal-use GDP modeler.

provide means for defining the shapes of components and sometimes allowable shape variations (tolerances), for positioning component representations to define assemblies, for calculating properties (appearance, mass, etc.), and—when linked to CAM modules—for generating manufacturing-process data such as NC programs.

One can discuss these systems in terms of the generic geometry system shown in Figure 2. Representations (models) of objects are built from definitional data supplied by users, and procedures are evoked by user commands to compute properties and do other useful work. The users may be humans, as is almost universally the case in design, or programs—increasingly the norm in manufacturing applications, where modeling systems are used as utilities by programs that simulate the motion of robots, check the correctness of NC programs, and so forth. The effectiveness of systems of the type shown in Figure 2 is set mainly by the intrinsic power of the internal representation schemes—what can be represented, and with what fidelity—and by the procedures that can be deployed to calculate useful results. Nearly all of the pre-1980 systems carried ambiguous representations that required human interpretation to be useful, whereas the new-generation solid modeling systems carry unambiguous representations that permit many calculations to be automated, at least in principle.

The Evolution of Computer Modeling

Figure 3 summarizes the evolution of modeling technology. The early roots can be traced to the 1950s, when computer graphics was invented, the first programming languages were devised for the then-new NC machine tools, and some computationally useful segments of projective geometry became popular in pockets of the engineering community. This early work nurtured the four largely independent streams shown in Figure 3, which are only now beginning to merge.

Wireframes

The wireframe stream supplied nearly all commercial CAD and CAM systems until the advent of commercial solid modeling. These systems appeared first as simple two-dimensional programs for designing printed circuit boards and digitizing mechanical drawings, one view at a time. In the 1970s the systems' modeling entities—two-dimensional lines and arcs—were generalized to represent segments of three-dimensional space curves that could be linked to represent the edges of solids (hence the name

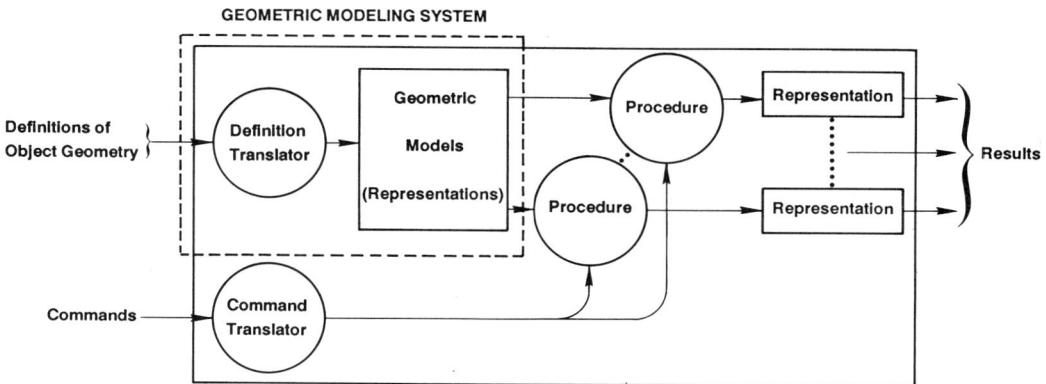

FIGURE 2 A generic geometric modeling system.

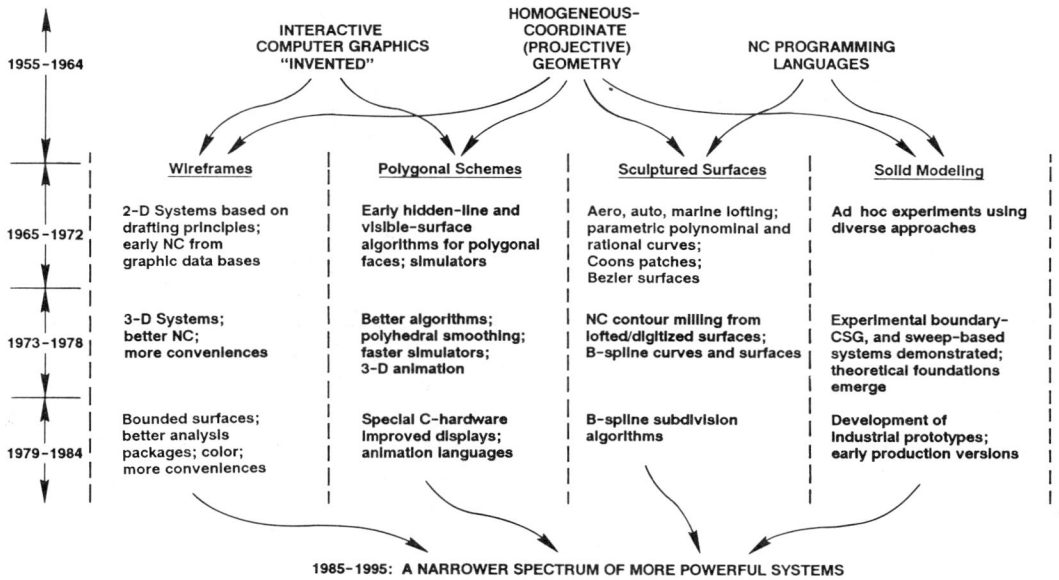

FIGURE 3   Evolution of mechanically oriented modeling technologies.

"wireframe"). Wireframe representations of solids can be projected computationally to generate multiple-view orthographic, isometric, and perspective drawings that, with human cuing to control visibility, mimic manually produced drawings.

Figure 4 demonstrates two serious deficiencies of wireframes: They may be ambiguous, and they may represent invalid ("impossible") solids. Specifically, three distinct solids exhibit the edges displayed in perspective on the left in Figure 4, whereas no solid can have the edges implied in the right-hand illustration. In principle both conditions can be detected automatically, but detection is computationally expensive and automatic repair is impossible. Thus, wireframes cannot be used as primary rep-

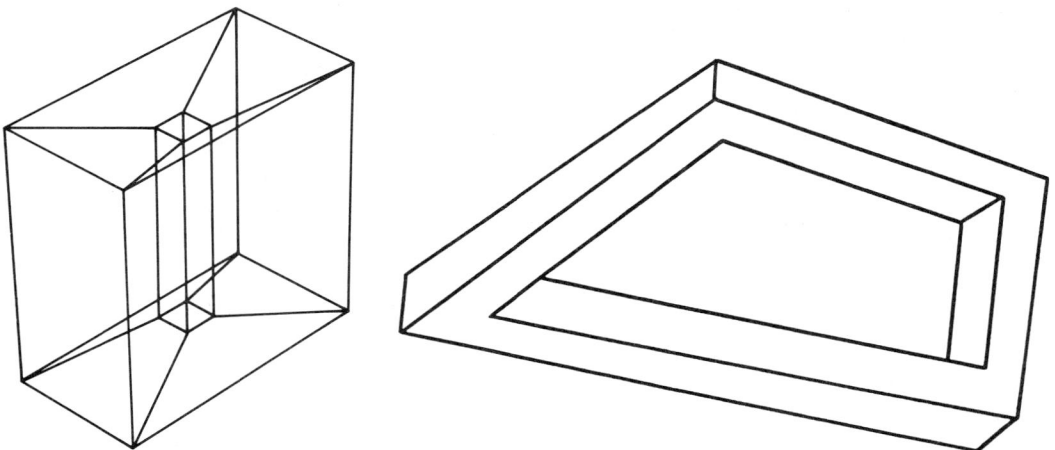

FIGURE 4   Deficiencies of wireframes: ambiguity (left) and invalidity (right).

resentations in automated systems. Nevertheless, wireframe modeling systems proved useful in the largely unautomated 1970s because they offered essentially paperless drafting as well as electronic data management for handling revisions. They continue to be useful today for somewhat different reasons, as will be noted later.

Solid Modeling

Solid modeling is distinguished by the use of valid and unambiguous representations of solids. It is the newest mechanical modeling technology and almost certainly will replace wireframe technology as various system problems (noted later) are resolved. Figure 5 shows the two schemes that are used most frequently: boundary representations (b-reps), in which solids are represented by sets of faces that enclose them completely, and constructive solid geometry (CSG), in which solids are represented as Boolean combinations (unions, differences, and intersections) of simple primitive solids.

Four other unambiguous schemes for representing solids are known and used, often in conjunction with boundary or CSG schemes, for certain kinds of applications:

• *Spatial Enumeration.* A solid is represented (usually approximated) as a union of quasi-disjoint box-shaped cells "filled with matter." The cells may be of uniform size or of varying sizes if generated by recursive binary spatial subdivision. Enumerations of the latter type may be organized as logical trees, called quadtrees in two dimensions and octrees in three dimensions.

• *Cell Decompositions.* A solid is again represented as a union of quasi-disjoint cells, but now each cell may have a distinctive shape, provided that it is homeomorphic to a sphere. Triangulations are the simplest form of cell decomposition, and finite-element meshes are the most widely used engineering embodiment.

• *Sweeping.* A solid is represented as the spatial region traversed ("swept-out") by either an area or a solid moving on a spatial trajectory. Although sweeping is central to modeling motional processes such as machining and robotic assembly, there are many open mathematical and computational questions surrounding it.

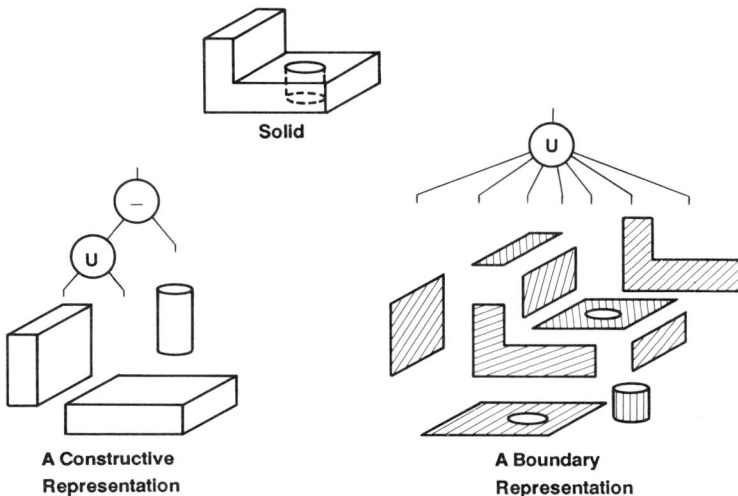

Solid

A Constructive Representation

A Boundary Representation

FIGURE 5 Solid modeling examples: constructive solid geometry and boundary representations.

• *Primitive Instancing.* This is a formalization of the family-of-parts concept. A solid is represented as a particular member of a family—say, the family of single-diameter round shafts with oil grooves—by supplying appropriate numerical parameters to a family-specific collection of formulas for displaying members of the family, calculating their mass properties, and so forth.

The roots of solid modeling can be traced to a few experimental systems built in the early 1960s that largely failed. The first successful experimental systems appeared in the early 1970s, mainly in European, Japanese, and American universities. Formal theories of solid modeling began to appear a few years later. In the late 1970s a second generation of experimental systems appeared. These seeded a first generation of vendor-supplied industrial solid modeling systems that appeared in the early 1980s.

Polygonal Schemes

The polygonal-scheme stream in Figure 3 could be retitled "graphic rendering" in that the goal is to provide visual effects. These effects range from real-time imagery for flight simulators, through commercial animation (as used in television, for example), to research in visual perception. This stream draws its title from the representation scheme that is common to all such applications—collections of polygons that approximate the boundaries of the objects being displayed. Extensive research has focused on developing fast algorithms and special computer hardware for generating displays from polygon lists, and some of this technology has been incorporated recently in industrial solid modeling systems to build approximate boundary representations.

Sculptured Surfaces

The sculptured-surfaces stream has the oldest roots, which lie in the mathematics of curves and surfaces. The first design applications appeared in the 1950s and 1960s, when Coons (1967), Bezier (1972), and a few other pioneers sought to replace the lofting and clay modeling techniques used in the aeronautical, marine, and automotive industries with computerized descriptions of doubly curved surfaces. Subsequently, there has been almost continuous development of mathematical bases and computer techniques for representing curves and surfaces, but until about 1980 little attention was paid to algorithms for processing surfaces: for example, computing curves of intersection or testing closedness to determine whether a surface may qualify as the boundary of a solid.

Contemporary Modeling Systems

Wireframe systems are at, or close to, the practical limits of their potential. They are still being installed in significant numbers because large collections of semiautomatic application codes are available for wireframe systems, there are large numbers of trained users in industry, many thousands of parts have been defined through wireframe systems, and a new generation of PC-based systems makes wireframe technology accessible to small firms. Polygonal systems have never played a major role in industrial modeling. Sculptured-surface systems tend to be proprietary within each major organization (e.g., airframe company) and are regarded as special-purpose systems rather than general mechanical modelers. Thus, the future lies with solid modeling, but, for reasons noted later, this modeling system is not yet ready to take over all of the modeling now done through wireframe and sculptured-surface systems.

System Organization and Geometric Coverage

The 1970s solid modelers fell into one of the two families shown in Figure 6. They

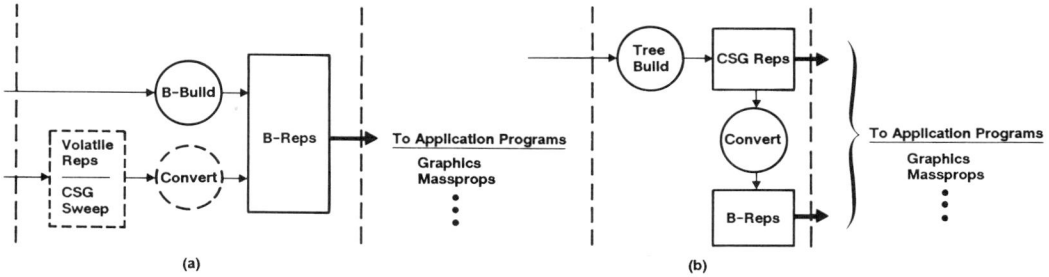

FIGURE 6   The single- and dual-representation architectures characteristic of the 1970s.

either had a single primary representation scheme, usually of boundary type, or dual (CSG, boundary) schemes, with the dual representation being computed from the boundary representation. Nearly all of the 1970s systems were quadric-surface modelers; that is, they could describe only objects bounded by first- and second-order surfaces—in practice, planar, cylindrical, spherical, and conical surfaces (the so-called natural quadrics). As we shall see later, the natural quadrics cover almost all unsculptured, functional mechanical parts.

The emergence of commercial solid modelers in the 1980s brought greater organizational variety. Figure 7 shows the trends: multiple representations,[2] some representations (at least one) being exact and the others approximate; auxiliary representations for several purposes, such as to speed up important algorithms and to carry attribute data; and a collection of geometric utilities available for use within the modeler and also by external applications. Thus, for example, many systems now compute planar approximations to curved surfaces using technology developed in the polygo-

nal stream (see Figure 3), and some systems maintain octree approximations. Representation-conversion algorithms are the "glue" that holds such systems together, and the maintenance of consistency over the whole set of representations when any one representation is edited (to install an engineering change, say) is a major system-design challenge.

An important current goal in solid modeling is installing "exact" sculptured-surface facilities to replace the planar-approximation methods used in some systems. This is proving to be considerably more difficult than expected, with the calculations needed to implement Boolean operations posing the main problems. Boolean operations are essential for many purposes, such as modeling material removal and detecting collisions

---

[2]An unambiguous representation is guaranteed to contain, in principle, enough information to allow any computable geometric property of the represented solid to be calculated automatically. This means that, in principle, a modeling system need contain only a single unambiguous representation scheme. In practice, however, no single scheme can support a range of applications efficiently, and hence the interest in multiple representations.

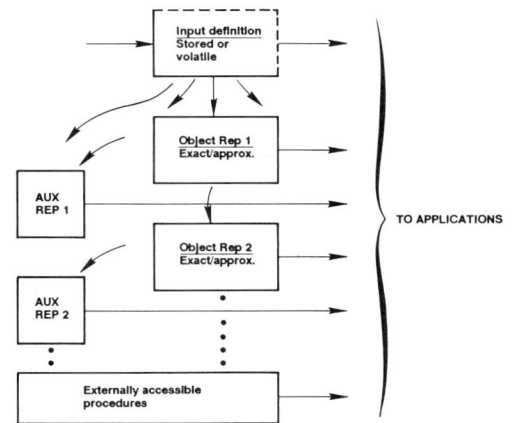

FIGURE 7   Generalized modeling system architecture.

between moving bodies and interference in assemblies, and as conveniences in defining parts; Boolean implementations, however, are mathematically delicate and computationally intensive.

Applications

Solid modeling has the potential to support the automation of almost all conventional technical tasks done in industry, from detailed strength analyses through graphic rendering to the automatic planning of machining and assembly operations and the programming of tools to do the work. We say this with confidence because we have mathematical proof that our representations of parts, fixtures, etc., are "informationally complete." This proof, however, is an existence proof; it tells what is possible without telling how to accomplish it. The fact that relatively few tasks are automated today is due primarily to our lack of scientific understanding of the tasks; succinctly, we have inadequate mathematical task and process models, and without these we cannot write reliable applications codes. A brief applications status report follows.

We wrote almost 5 years ago (Requicha and Voelcker, 1983, 29–30) "only three major applications—graphics, mass properties [volume, centroid, inertia tensor], and static interference checking—are understood well enough to be handled automatically in most systems. . . . Modelers that can support only [these applications] are difficult to justify in most industrial installations, and thus vendors are using existing packages to provide numerical control and other services while awaiting more advanced modules that can exploit the power of solid modeling." There has been little overt change in the intervening years, and thus Figure 8, taken from the 1983 reference, depicts the current situation reasonably accurately: a few applications are handled automatically from the solid modeler, and the others through human-interactive programs devised mainly for wireframe modelers. The requisite wireframe representations are easy to derive automatically and download from the solid modeler; uploading from a wireframe modeler to a solid modeler requires considerable human assistance.

Although there has been little overt change in the automation of applications over the past several years, applications research has progressed steadily, and one can expect a few of the dashed lines in Figure 8 to become solid by 1990, with most of the

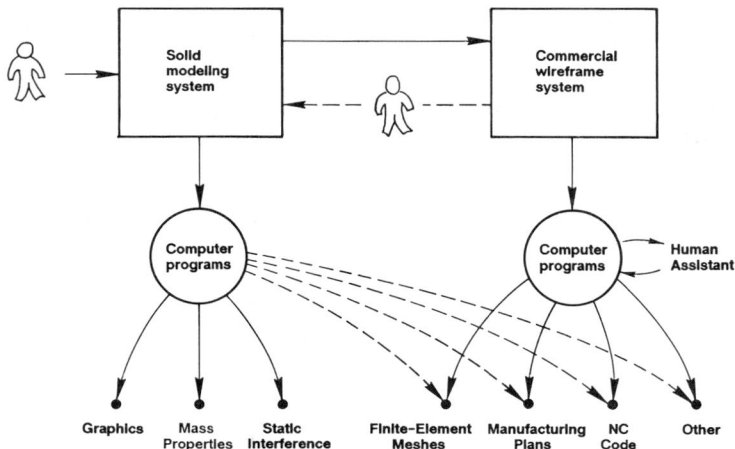

FIGURE 8   A contemporary, and probably temporary, marriage of convenience.

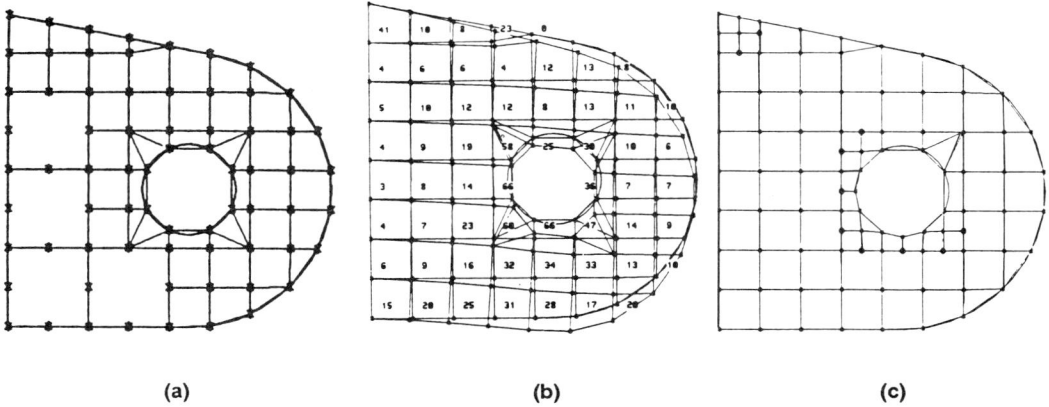

(a)   (b)   (c)

FIGURE 9   Two-dimensional finite-element meshes: (a) mesh in two dimensions; (b) loaded mesh; (c) automatically refined version of mesh.

others (including several not shown) following before the end of the century. Two examples of current research will be discussed, together with summary comments on other applications.

• *Automatic finite-element analysis.* Automatic finite-element mesh generation has been under study since the late 1970s, and industrially viable automatic mesh generators can be expected by 1990. There are several approaches to the problem, with one of the most promising being a two-stage process using quadtree or octree enumeration to mesh the interior of a solid, followed by boundary traversal to extend the interior mesh to the surface of the part (Kela et al., 1986; Kela, 1987). Figure 9a shows such a mesh in two dimensions, with the interior quadtree structure of graded blocks clearly

visible. One could stop at this point and submit the mesh to a standard analysis program such as NASTRAN; however, there is much to be gained by treating mesh generation and mesh analysis as coupled problems. This is accomplished by including a correction loop, as shown in Figure 10. Thus, errors associated with the analysis, such as high stress gradients, can be used to guide local refinement of the mesh and subsequent incremental reanalysis. Figure 9b shows the loaded mesh (the left face is fixed and the hole carries a downward force), with error indicators written in each element. Figure 9c shows an automatically refined version of the mesh.

The solid modeler shown in Figure 10 supplies part geometry and, through an attribute facility, loading and boundary conditions. The modeler also generates the

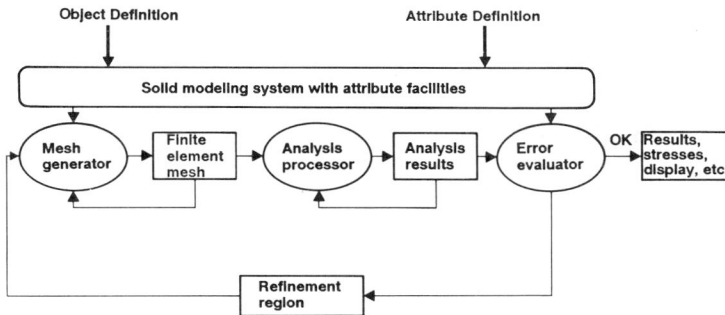

FIGURE 10   An automatic finite-element analysis system.

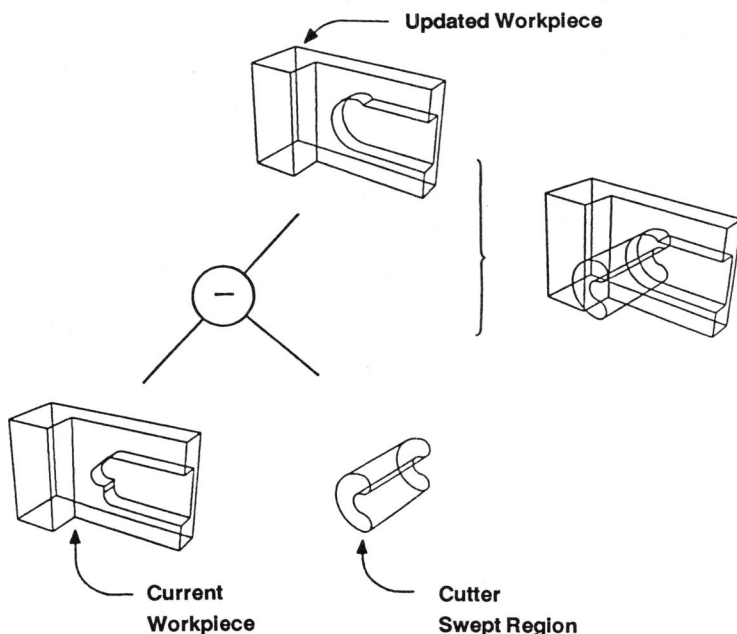

FIGURE 11   The basis of machining simulation.

quadtree or octree approximations used in the meshing procedure, plus other aids for managing the process. A two-dimensional version of the system shown in Figure 10 is running (Kela, 1987), and an experimental three-dimensional version can be expected in 1 to 2 years.

• *NC machining.* Figure 11 shows the essence of machining simulation, which is easy to do given a solid modeler with appropriate geometric power. The driving relation is $W_i = W_{i-1} - V_i$, where W_i is the workpiece after (simulated) execution of the ith NC command, V_i is the spatial region swept by the cutter on the ith command, and "$-$" is the (regularized) set-difference operator. Thus, a simple simulator reads an NC program block by block and displays the workpiece after each command; a person watches the displays and tries to spot problems (collisions, invasive machining, etc.). Machining simulators of this form can be expected soon for industrial applications.

Automatic NC-program verification[3] seeks to do two things: detect problems without recourse to human observers, and determine automatically whether the final machined part W_F is identical to the desired part P (Sungurtekin and Voelcker, 1986). The latter goal-attainment test ($W_F = P$?) is easy to do in a solid modeler in principle, but there are computational subtleties. Automatic problem detection is done by applying two different kinds of tests at each stage of a simulation. Spatial problems are detected by various intersection tests, with $P \cap V_i = \emptyset$? (does the current cutter-swept region V_i intersect the desired part P?) being the test for invasive machining. Figure 12 shows output from a verifier when a cutter executing a positioning (rapid) motion collides with the workpiece and fixture. The detection of "technological" problems, such as cutter breakage or

[3]The terminology in the field is not standardized. Some authors use "verification" to denote what we have called "simulation."

violation of tolerance constraints, mainly requires force calculations that are done indirectly. For example, $R_i = W_{i-1} \cap V_i$ is the "solid" actually removed (made into chips) by the ith command, and the volume of R_i can be calculated automatically by a modeler's mass-property module. From this and the known cutting conditions, such as path length, and feed rate, an average material removal rate can be calculated. From the removal rate and other data it is possible to estimate the average forces on the cutter and hence predict its deflection, breakage, and so forth. Methods for estimating peak forces through higher moments of R_i are under study.

Although at least one NC verifier program should be ready for industrial tests in a year or so, automatic NC program generators are farther away. High-level machining programming languages coupled to solid modelers should appear by 1990 (Chan and Voelcker, 1986). These programs probably will be followed sometime in the 1990s by automatic process planners with outputs of programs in high-level machining programming languages.

• *Other applications.* Industrial robotics clearly offer fertile ground for model-based automation, with off-line manual and automatic programming being prime targets. The situation here is similar to that in machining: Graphic simulators are essentially available and automatic program verifiers and high-level languages should be in use by about 1990, but automatic planning and programming lie farther away. Another fertile problem area is simulating material flow in dies, as in injection molding and forging, and progressing from simulation to computer-aided and computer-automated

GO – RAPID

***** **ERROR: COLLISION WITH THE WORKPIECE**

***** **ERROR: COLLISION WITH THE FIXTURE**

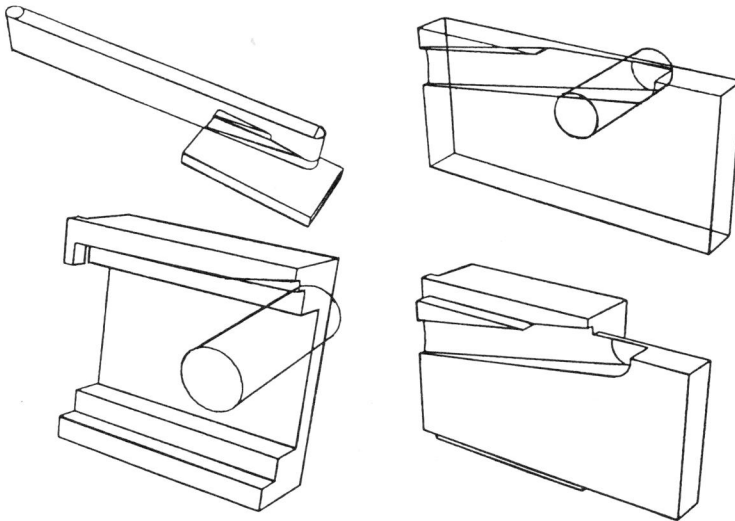

FIGURE 12   Automatic detection of collisions in an NC program verifier.

die design. This work has a high payoff potential but low public visibility, because the underlying physical processes are complicated (Dawson, 1986). Solid-modeling-based flow simulators are already in use in a few companies (Wang et al., 1986).

Although it does not exhaust the list of applications, the foregoing survey should convey a sense of the current state of the art. Some issues close to part and product design are discussed later.

User Interfaces

Early CAD/CAM systems were designed to be electronic drafting boards. T square, compass, and triangle were replaced with pointing devices (cursor, light pen, etc.) and command menus whereby users could create lines, circles, arcs, free-form curves, icons (e.g., arrows), and text. Users could establish relations between elements of a drawing, for example, making one element parallel, perpendicular, or tangent to another and could copy, rotate, translate, save, and delete entities. These drafting interfaces came to be highly engineered, convenient, and fast as computer-graphic technology advanced, but they enforced almost no model-based discipline on the user. These systems could be used to draw anything, because there were no underlying mathematical models of any object of higher order than curves.

When wireframe systems appeared, drafting interfaces generally were extended, rather than redesigned, to exploit the mathematical rules governing wireframe structures. Thus, users continued to work mainly in orthographic views, with some behind-the-scenes view-linking, and they retained much of the draftsman's credo that one drafts to communicate shape to humans rather than to define mathematically correct models of solids.

The advent of solid modeling forced serious thought to be given to the design of user interfaces, beginning about 1980, for several reasons. First, many solid modelers emerged from the research laboratories with command language interfaces rather than graphic interfaces; thus, there was interface design to do, since engineers often resist "programming" and insist on graphics. Second, solid geometry is usually created in chunks—whole blocks and cylinders—rather than through lower-order lines and arcs. Thus, the highly engineered drafting interfaces became largely irrelevant. Finally, solid modeling requires three-dimensional thinking and visualization skills; thus, three-dimensional displays (perspective line drawings and shaded images) are almost essential, because defining entities in three dimensions is more difficult than in two dimensions, and working through two-dimensional views is often not the best approach.

Most contemporary solid modelers have "first-generation" solid-oriented graphic interfaces.[4] These are, in essence, graphic versions of simple command languages that permit primitive solids to be instantiated from menus, positioned through rigid motions and coordinate-system declarations, and combined through Boolean operations. Figure 13 shows some of the menus and displays in a typical system (McDonnell Douglas's UNISOLIDS™). Many systems also provide means for "extruding" and "swinging" closed planar contours into translationally or rotationally symmetric solids, and the newest interfaces (e.g., version 4.0 of UNISOLIDS™) offer simple "features" such as countersunk holes and various kinds of slots and pockets as definitional primitives. Few contemporary modelers offer relational facilities that would, for example, allow a user to "Put face A of solid B against face C of solid D," and none, insofar as the author knows, supports con-

---

[4]A few systems, e.g., MEDUSA™, were designed from the outset with drafting-like input facilities in mind, but these required internal compromises that may limit the systems' longevity.

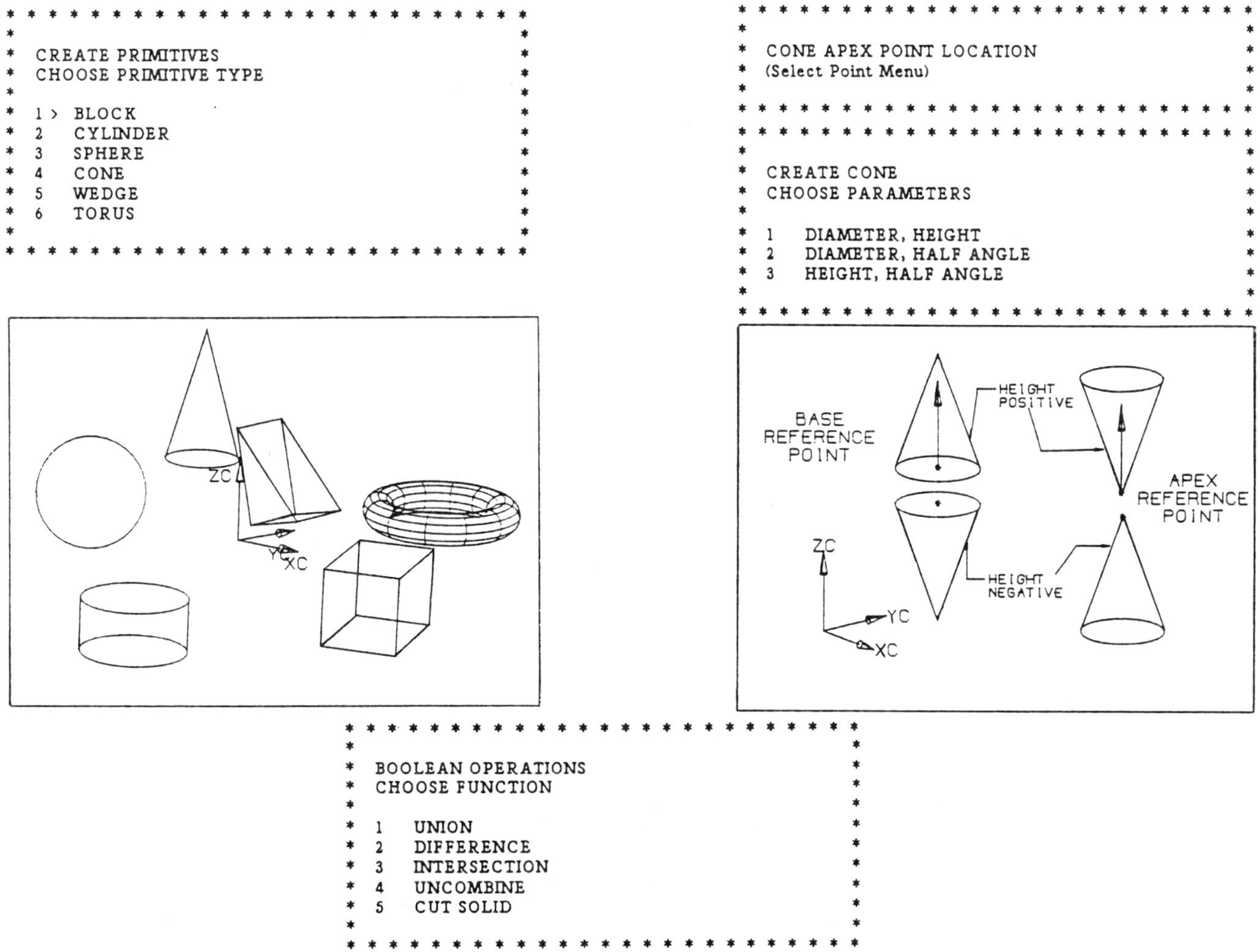
CREATE PRIMITIVES
CHOOSE PRIMITIVE TYPE
1 > BLOCK
2 CYLINDER
3 SPHERE
4 CONE
5 WEDGE
6 TORUS
ZC
YC
XC
CONE APEX POINT LOCATION
(Select Point Menu)
CREATE CONE
CHOOSE PARAMETERS
1 DIAMETER, HEIGHT
2 DIAMETER, HALF ANGLE
3 HEIGHT, HALF ANGLE
BASE REFERENCE POINT
HEIGHT POSITIVE
APEX REFERENCE POINT
HEIGHT NEGATIVE
ZC
YC
XC
BOOLEAN OPERATIONS
CHOOSE FUNCTION
1 UNION
2 DIFFERENCE
3 INTERSECTION
4 UNCOMBINE
5 CUT SOLID

FIGURE 13 Menu-driven solid modeling.

strained design, wherein critical parameters of parts are found automatically by solving systems of equations. The appearance of such facilities will mark the transition into second-generation interfaces.

Thus far we have focused on graphic interfaces for human users, but, as we noted earlier, automata (programs for automatic finite-element analysis, machining simulation, and so forth) also use modeling systems; indeed, programs are likely to be the major users within a decade. Formal languages are the appropriate interfaces for automata and also for humans who wish to design parametrically. Languages are becoming highly developed for modelers with CSG input facilities (because representational validity is easy to guarantee in CSG).

The following provides a simple example wherein a generic shelf is defined as an assembly of a board (a solid-block primitive) and two brackets defined separately as generics:[5]

GENERIC SHELF (SHELFPART);
PARAM LEN {New length value for
    bracket};
PARAM T {New thickness value for
    bracket};
LEN = 6; T = 0.5;
BOARD = BLO(X=24, Y=0.75,
    Z=8);
BRACKET1 = BRACKET(L=LEN,
    T=T) AT MOVX=3;
BRACKET2 = BRACKET(L=LEN,
    T=T) AT MOVX=20;
SHELFPART = BOARD ASB
    BRACKET1 ASB BRACKET2.

Note that some parameter values in this definition are fixed whereas others are variables with default values. This language allows algebraic and trigonometric expressions and conditional statements but not recursion or iteration.

In summary, two quite different modeler

interfaces are available. Definitional languages offer much of the abstractive power of programming languages—naming, scoping, conditionals, and so forth—and are the natural input media for automata, but they do not exploit humans' visualization skills. Interactive graphic interfaces are powerful aids to human spatial reasoning but lack abstractive power. (It is difficult, for example, to do parametric design through a graphic interface.) A new medium combining the strengths of graphics and definitional languages is needed, but no serious candidates are on the horizon.

Current Limitations and Research Frontiers

Computer modeling and its applications in the mechanical industries have burgeoned in the past 20 years. We have come a long way, but we still have a long way to go. The major current limitations of modeling and applications technology are summarized next, and the state of pertinent research is discussed.

Solid Modeling

In the following we assume that solid modeling will become the dominant medium for describing parts and products (and stock, fixtures, etc.) in the mechanical industries and that it will subsume useful techniques from the other streams shown in Figure 3.

Theory. Current theory, most of which is less than a decade old, can be thought of as having three components: an existential (representation-free) theory of solids as point-sets or spatially embeddable topological polyhedra,[6] a companion theory of rep-

---

[5]"Generic" is PADL (Part and Assembly Description Language) jargon for an archived parametric definition (Hartquist and Marisa, 1985).

[6]There are two competing existential theories, one based on "r-sets" (compact, regular, semianalytic sets in E^3) and one based on manifolds. The differences, and their practical implications, are too subtle to delineate here.

resentation (the six unambiguous schemes summarized earlier), and a growing collection of functions and algorithms that are central to building and maintaining representations and to converting between representations. This body of theory covers the nominal or ideal-form geometry of rigid solids, and the qualifiers "ideal-form" and "rigid" mark the main limitations:

• The ideal-form restriction limits severely the ability to handle tolerancing—i.e., allowable variations in form, position, and relation (to other entities). The issues here are complex and subtle (Requicha, 1984). Although many applications can be automated without a satisfactory theory and a technology of tolerancing, others—automatic planning of machining, automatic die design, automatic tolerance assignment in assembly design, and a few others—probably cannot. A growing volume of research is focused on tolerancing.

• The rigidity restriction can be discussed for present purposes in terms of objects that are nominally rigid and objects that are not. Linear deformations of nominally rigid solids—for example, elastic strain and thermal expansion—can be handled relatively easily by finite-element methods. Mass-preserving plastic deformation, as in forging and extrusion, can be handled similarly if the constitutive relations are known. Certain non-mass-preserving deformations, notably machining, can be handled easily by different methods, as indicated earlier. The really open issues lie with objects that are not nominally rigid. These include nominally elastic objects (e.g., snap fasteners), flexible objects (e.g., electric-cable harnesses), and limp objects (e.g., upholstery and padding materials). No systematic means, much less unified means, exist for modeling such objects, and thus far little research has been focused on them.

System Technology. Contemporary industrial solid modeling systems operate under the nominal-form and rigidity restrictions just cited. Their industrial usefulness is limited by the following additional factors (as well as other factors associated with applications, which are discussed separately).

Adequately robust systems have limited geometric domains—typically objects bounded by the natural quadric surfaces. As noted earlier, efforts to incorporate sculptured surfaces are encountering problems, but—given the talent being focused on domain extensions—these problems are likely to be solved in one way or another in the next several years.

All systems, except for a few optimized for special applications, are slow because solid modeling is computationally intensive. This problem is also attracting talent and resources and in the next several years will be ameliorated (it will never be solved to the full satisfaction of users) by better algorithms and massive computing power—initially supercomputers for critical applications and, within 5 years, board-level hardware accelerators optimized for solid modeling (Goldfeather and Fuchs, 1986; Kedem and Ellis, 1984).

All systems, except for a few optimized for special applications, seem to be subject to a complexity barrier that limits their effective domains to objects representable by about 10^3 or fewer representational primitives (faces, primitive solids, etc.).[7] This limit was acceptable 5 years ago with computers capable of only about 1 million inferences per second. But computers 10 times faster and 10 times larger (i.e., 10 times more on-line storage) are becoming available, and users would like to handle objects 10 times larger now and objects 100 times larger 5 to 10 years hence. Current modeling software probably will not support such

---

[7]By way of contrast, automobile engine blocks require $O(10^4)$ representational primitives and whole-engine assemblies require $O(10^5)$. Current systems can handle such objects in principle, but in practice the costs, delays, and technical difficulties of doing so are almost always unacceptable.

growth because it treats all aspects of a definition as equally important, from the smallest hole to the largest macrofeature. Humans cope with complexity through abstraction and "dynamic hierarchies," which enable them to ignore two kinds of information—that which is irrelevant and that which is too detailed. Modeling software must be endowed with similar faculties, and research toward this end is in progress.[8]

Applications of Solid Modeling in Manufacturing

Five years ago manufacturing-process automation was concentrated at the effectors (e.g., at the machine tools), and the requisite upstream support in the form of manual process planning, machine-tool and robot programming, etc., was expensive unless production runs were long. Today, model-based automation of some of the upstream activities is imminent, as discussed earlier, and almost complete "vertical automation" of some important manufacturing processes seems attainable before the end of the century. The key to "vertical" manufacturing-process automation seems to lie in finding effective computational models for processes (machining, forging, dextrous assembly, etc.). A century of classical research on processes, when coupled to new solid-modeling and analytical tools, provides a strong springboard into a new world of computational process planning and process control. Thus, modern manufacturing research is growing steadily in both breadth and depth, and dramatic progress seems likely in some areas.[9]

Applications of Solid Modeling in Design

The status of modeling in design, which is the subject of this paper, is a slippery topic, in part because "design" is both a noun and a verb. We shall approach the subject cautiously, proceeding from the evident to the conjectural.

Design Definition. In current industrial practice, a finished design (noun) for a product is defined by four coupled bodies of information:

1. Ideal-form (shape) specifications for the component parts
2. Associated variational specifications (tolerances)
(Note that these first two items taken together are equivalent to "detail drawings.")
3. Component-combination specifications ("assembly drawings")
4. Material and finish specifications

Performance specifications rank as collateral information or as part of the design-process documentation; they cannot be part of the design definition unless consistency with the four components can be guaranteed, in which case performance specifications are redundant (because they are derivable from the design definition, at least in principle). Note also that manufacturing- and assembly-process specifications are not included in the design definition—a matter we shall discuss later.

Contemporary object modeling theory and technology can handle items 1, 3, and probably 4, at least in principle (i.e., subject to the geometric coverage, complexity, etc., limits already noted.). Item 2—tolerances—is a problem area, as noted earlier.

---

[8]The underlying problem is that solid modeling algorithms run in polynomial rather than linear time, and thus a tenfold expansion of a model may require a fiftyfold expansion of computing power to maintain a given level of performance. One way to avoid this is to treat only the "currently relevant" portions of the larger model.

[9]We note for completeness that automatic operation of manufacturing systems—machining cells,

---

CIM complexes, whole factories—involves different lines of research into different types of problems (scheduling, line balancing, etc.) using different models (probabilistic discrete-event models, rather than mainly deterministic Euclidean and Newtonian models). This is the province of industrial engineering and operations research.

Design Validation. Physical testing provides the ultimate validation for a design by ensuring that the product is, for example, strong enough or not too heavy to meet its performance specifications. But physical testing is expensive and time-consuming, and therefore computational analyses are being substituted as confidence in modern analytical methods grows. Contemporary modeling systems already support automatic mass-property calculation and soon will support automatic finite-element and more specialized kinematic and kinetic analysis procedures. Thus, computational design validation is in relatively good shape because few of the pertinent analytical procedures require tolerancing information.

Optimization of Parametric Designs. As a design approaches completion, a stage is reached where the design is explicitly or implicitly parametric. That is, ranges for a small number of key parameters are known, and one seeks values for these parameters that optimize the design under metric criteria. (A typical problem is minimizing the weight of a critical component under strength constraints, with certain shape parameters fixed and others variable.) Here again the analytical procedures supported by modern parametric object modelers are adequate, at least in principle, if coupled to optimization software (gradient-driven hill-climbing algorithms, linear-program solvers, and the like).

Support of Conceptual and Preparametric Design. Some issues that arise can be posed through the simple example shown in Figure 14. A designer begins the design of a simple component with three holes of known diameter and configuration (Figure 14a); these mate with features of other parts. The designer then creates some bosses to contain the holes (Figure 14b) because of concern about interference with other components passing between the holes. Finally, the holes and bosses are bound together into

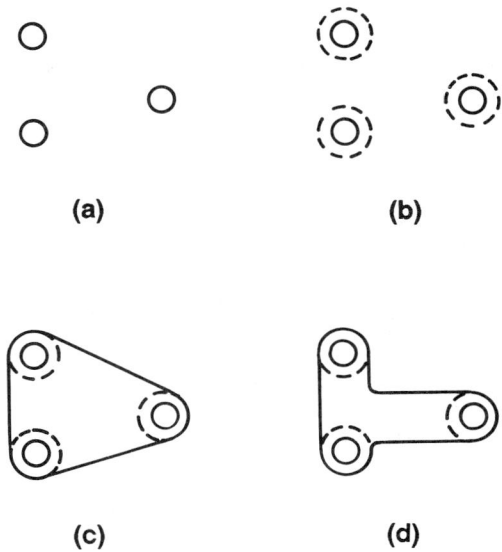

FIGURE 14   Design of a simple component.

a single part as in Figures 14c and 14d, with the final shape being governed by criteria for clearance, strength, weight, and simplicity.

The process described is easy to do in an electronic drafting system or in most wireframe modelers, but of course little can be calculated along the way because such systems are mainly display engines. The process is not easy to do, as described, in many contemporary solid modelers; one must "trick" these systems—for example, by defining the holes as an assembly of solid cylinders that will be "subtracted" later, when the encompassing solid has been defined. The situation gets worse when one tries to do preliminary assembly design in a solid modeler. Broadly, most current modelers require that all entities be valid attributes of well-defined solids, and, if properties are to be calculated or displays generated, all parametric variables must be assigned values.

In summary, it is fair to say that contemporary solid modelers are suitable for documenting (creating definitions of) finished designs and for supporting the calculations needed for design validation and parametric optimization of nearly finished designs. Current systems provide little help during (and indeed may hinder) the conceptual and preparametric stages of design (as done by humans). There is a growing body of research, much of it based on artificial intelligence, aimed at this problem.

This is not the end of the story about modeling in design, however, because thus far we have addressed only the modeling of objects. By taking a broader view, we shall see that other types of modeling are equally important.

A Broader View of Modeling in Design and Manufacturing

J. R. Rinderle (1986, 1987) has suggested a triad of function, form, and fabrication as a mechanism for discussing the design and manufacture of mechanical products. The following paraphrases seem to capture Rinderle's notions:

• *Form* refers to the product as a physical artifact (typically an assembly of sub-artifacts) having shape and various shape- and material-determined motional, thermal, and other characteristics.

• *Function* refers to what the product is

intended to do and can do, as contrasted with what it is (a physical entity).

• *Fabrication* covers the possible and actual means used to produce the product, and the methods used to mediate between alternative means.

If we associate the three terms with artifacts (call them "specifications" for present purposes) linked by transformations, we are led to the view shown in Figure 15, which extracts a portion of Figure 1: Form is induced from function through design, and fabrication is induced from form by manufacturing planning. Usually there is feedback (the dashed lines in Figures 1 and 15), with production planners recommending design changes to promote production efficiencies. Let us focus initially on the right half of Figure 15, because it is better understood than the left half.

Form artifacts are what we call designs (design definitions, specifications of designs); they consist of the four entities listed earlier—shape specifications, tolerances, and so forth—and we know how to model and represent these entities (except perhaps for tolerances). Fabrication specifications govern the manufacture, inspection, and assembly of parts and products; they consist of machining process plans, NC programs,

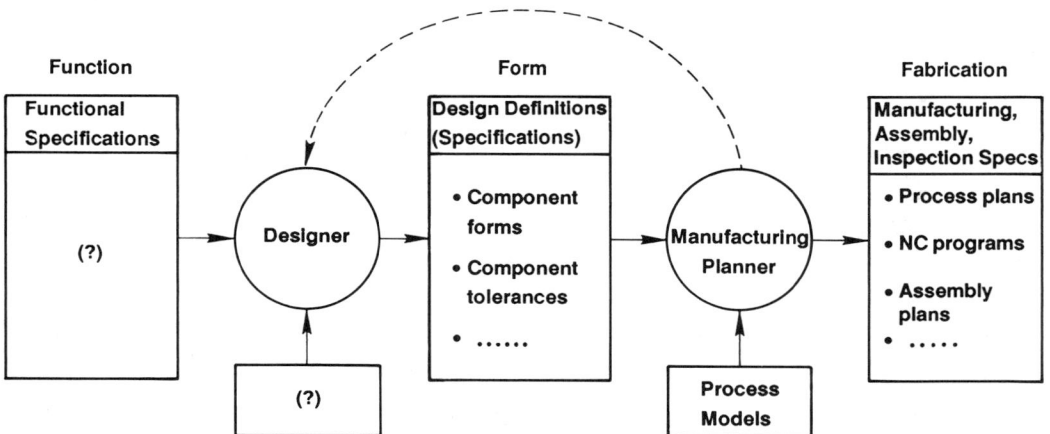

FIGURE 15   Form induced from function by design, and fabrication induced from form by manufacturing planning.

inspection plans, inspection-machine programs, etc. We are learning how to represent these specifications formally and model their effects mathematically and computationally. (See, for example, Chan [1987] and the earlier discussion of NC machining in this paper.)

Fabrication specifications are induced from form specifications by the manufacturing planner, whose main knowledge resources are sets of models for manufacturing and assembly processes, plus rules and procedures for selecting and sequencing processes. What is the nature of these models, rules, and procedures, and how does the planner use them? Here is a capsule summary of one view of these matters from the realm of machining, which is a relatively well understood manufacturing process.

One version of a machining-process model for use in machining planning is a function

(cutter, position, orientation, feed-motion)
→ {(surface-subset, position,
            orientation, precision)}

that defines the machined surfaces that are produced by a specific cutter fully engaged with a workpiece and fed in a specified manner. Thus, for example, an end-mill fed on a linear trajectory normal to its axis can produce two planar-surface subsets parallel to each other and the cutter axis, a cylindrical-surface subset parallel to the axis, and a planar-surface subset normal to the axis. Precision parameters are associated with each surface-subset to distinguish between nominally indistinguishable processes—e.g., boring and reaming. An analogous assembly-process model maps pairs of surface features on parts, a relative motion, and terminal conditions into "mated-feature" pairs with constraints such as a screw fully engaged in a threaded hole with a specified strain-induced residual torque.

One family of machining-planning strategies now under study uses inverted forms of these process models, namely,

(surface-subset, position, orientation,
precision)
→ {(cutter, position, orientation,
                    feed-motion)}

to establish, for a surface-feature to be produced by machining, a set of candidate cutters, setups, and feed motions. If a candidate set is established for every feature to be machined, then in principle, a machining plan can be constructed by combinatorial optimization over the candidate sets, using the verification tests (for invasive machining, etc.) discussed earlier to reject many candidate sets and ranking the survivors through such criteria as setup- and cutter-change minimization. Machining planning can thus be viewed as an iterative selection, instantiation, and sequencing of processes represented by inverted process models and subject to preconditions (e.g., on accessibility) and sometimes to postconditions.

Let us summarize the uses of modeling in the right half of Figure 15. We have models of solids under Form, and under Fabrication we have process specifications whose effects we can model (as in machining simulation). We also have, for use by the planner, low-level models that associate processes with "features" of solids. For machining, these are the forward (process → {surface-subset}) models and the backward (surface-subset → {process}) models.

In the left half of Figure 15 we have, under Form, models of solids as the "output" of design, but we have no models for Function, we have no intermediary models to aid the designer in associating "components of function" with "form-elements," and of course we have no procedures for aggregating "form-elements" into designed solids. Thus, the left half of Figure 15 is largely open.

TWO (OF MANY) OPEN ISSUES

Although it is tempting to speculate broadly about the mechanisms underlying design synthesis, we shall conclude by addressing briefly two issues that can be framed with enough precision to generate experimentally testable hypotheses.

Do Designers Have Too Much Geometric Freedom?

Let us begin with some data. In 1974–1975 a careful survey was made of the geometrical characteristics of the functional parts in a Xerox tabletop copier (Samuel et al., 1976). The primary purpose of the survey was to generate data on part geometry to guide the design of languages and processors in the PADL (Part and Assembly Description Language) family of CSG-based modeling systems (Brown, 1982; Voelcker et al., 1978). An important secondary purpose was to gather data on the geometrical characteristics of industrial parts as an end in itself, because we had been unable to find any such data before the survey.

The survey was conducted mainly by an experienced Xerox tooling engineer who devoted about 1 man-year to the task. He used the following versions of the PADL language as a meta-medium for survey purposes (the versions are distinguished by their sets of primitive solids; each is assumed to have general regularized Boolean operators and rigid-motion operators—translations for versions 1.n, translations and notations for versions 2.n):

1.0—Orthogonal Blocks and Cylinders

1.4—Orthogonal Blocks, Cylinders, and Wedges

1.6—Orthogonal Blocks, Cylinders, Wedges, and Cones

2.0—Orientable Blocks and Cylinders

2.8—Orientable Blocks, Cylinders, Cones, Spheres, and Tori

In essence, he assessed each part in the sample in terms of its describability in the language and assessed also the size of the resulting PADL definition.

Figure 16 shows some of the results.[10] The abscissa of Figure 16 shows primitive-instance counts: these are a close measure of the size of a CSG definition. The right-hand ordinate shows the percentages of surveyed parts that are describable in various versions of the language. For example, about 30 percent of the parts are describable in PADL-1.0 (the orthogonal blocks and cylinders version), with the largest PADL-1.0 definition requiring about 45 primitive instances; about 99 percent of the parts are describable in PADL-2.8, with the largest definition (that for the copier's base plate) requiring some 500 primitive instances.

The curve in Figure 16 labeled "Redesigned to PADL-1.0" is our present focus. This curve shows that 60 to 65 percent of the parts can be designed (or redesigned) in PADL-1.0 under criteria dictating that the parts be true functional and physical replacements (Samuel et al., 1976). Figures 17 and 18 provide a part-redesign example, with Figure 17 being the original and Figure 18 the redesigned part. If the copier had been designed from the outset with simple (PADL-1.0) geometry as an important design goal, the percentage of PADL-1.0 parts would have been considerably higher than 60 to 65 percent.

One can argue on various grounds that PADL-1.0 parts are technically and economically preferable to higher-version parts (Samuel et al., 1976), but here we conjecture merely that designers' "geometric freedom" can be restricted (put differently, designers can be subjected to "geometric discipline") with little or no loss in their ability to meet functional requirements. We believe that this conjecture warrants exper-

---

[10]Interested readers should consult Samuel et al. (1976) to understand the methodology and the underlying assumptions, which are important, and also to assess the many results not noted here.

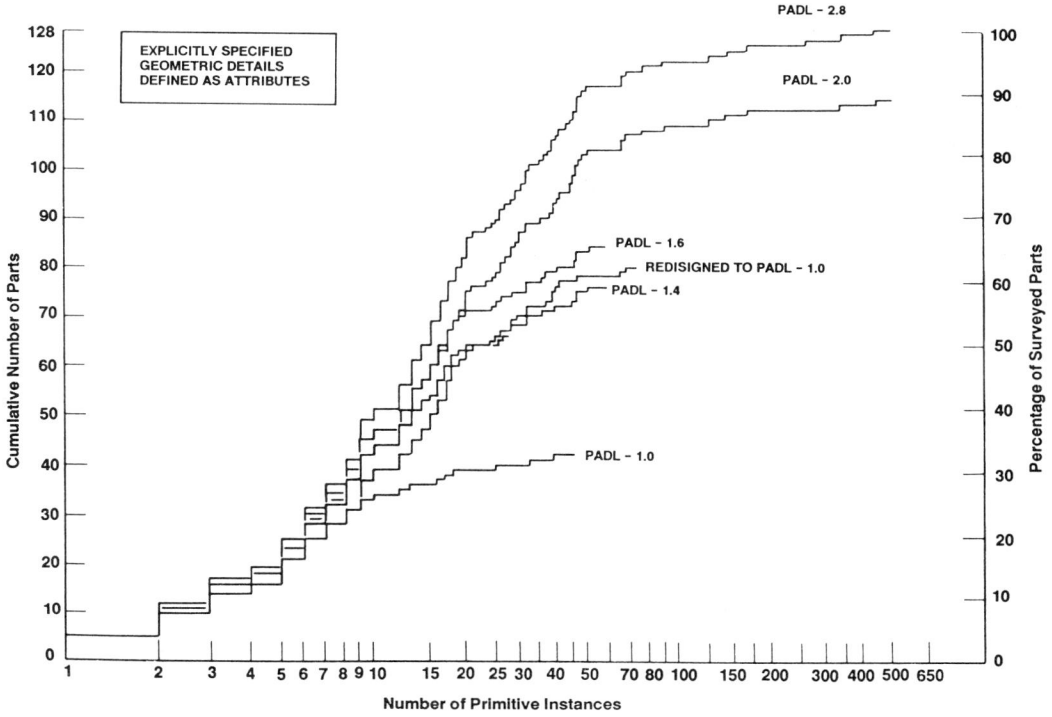

FIGURE 16   Some results from the Xerox part survey.

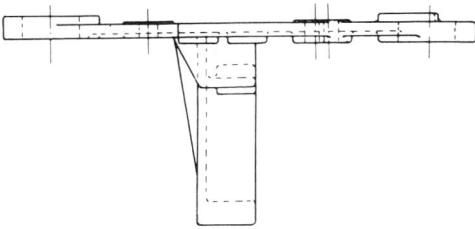

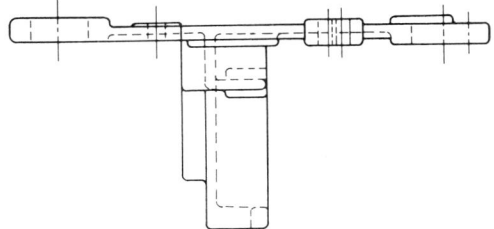

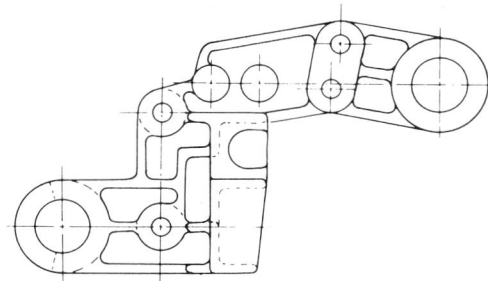

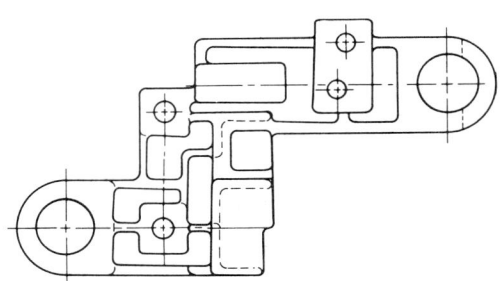

FIGURE 17   A production part as originally designed.

FIGURE 18   A redesign of the production part of
Figure 17, using PADL-1.0 primitives.

187

imental testing in industry and that experiments can be designed that will yield near-term practical benefits as well as longer-term insight into the nature of design.

Is Conditional Process Planning Preferable to "Open-Loop" Process Planning?

In current industrial practice the manufacturing processes needed to produce a part, and their sequencing, are wholly prespecified, and deviations are not allowed. This can be termed "open-loop" planning. If done correctly, it ensures that a correct part will be produced if each prespecified step is done correctly and that a part can be rejected if any step fails. Open-loop planning can be viewed as a factory- or production-organizing principle. It apparently arose from the master-gauge practices of the past century, and it had a subtle but profound effect on early tolerancing practice. (Or perhaps one should say that gauging and tolerancing practices have had a profound effect on manufacturing-planning and factory-organizing principles.)

Figure 19 shows a simple limit-tolerancing specification in which C = A + B (Requicha, 1977). This specification may be precisely what a designer needs to meet functional requirements efficiently, but it is not acceptable under current practice because the part cannot be made to the given specification by open-loop manufacturing

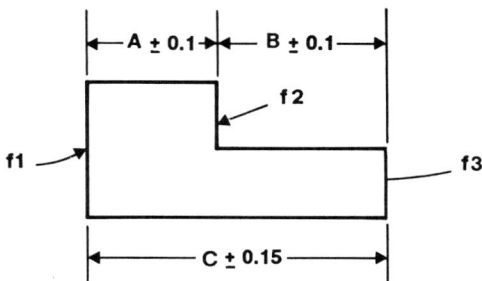

FIGURE 19 An "over-dimensioned" (or "over-toleranced") part.

methods. (Figure 19 would be termed "over-dimensioned" by almost all drafting supervisors.) To see why it cannot be made "open-loop," suppose that face f1 is produced first and used as a datum for the machining of f2 and f3. Face f2 is then machined, using a process with only enough precision to meet the f1,f2 tolerance of 0.1. If a planner can determine by measurement that the resulting f2 physical face lies near the center of its tolerance band, then a 0.1-process can be used to produce f3. If f2 lies near the edge of its band, however, a more precise process must be used for f3 to meet the f1,f3 tolerance of 0.15.

This simple example suggests several interesting questions, all researchable. For example, are designers capable of generating graph-structured tolerance relations, as in Figure 19, rather than the simple tolerance trees of traditional practice, and do such graphs really capture important functional relations? Are there families of useful manufacturing processes in which precision can be traded against cost, and are the trade-offs well understood?

One might finish with a third issue that follows from the above, namely, should the doctrine of interchangeability be examined to see if its benefits continue to outweigh its costs? For reasons of brevity we shall not pursue this topic and will close with the observation that interchangeability doctrines are already changing. Specifically, many modern products have a "replacement-module level" that is well above the single-part level and is rising steadily. For example, when the water pump in a car fails, the whole water pump is replaced rather than the one or two defective components of the pump. This means that the manufacture of artifacts that lie below the replacement-module level need not obey interchangeability criteria, but it is not clear that the implications of this new condition are being explored and exploited systematically.

SUMMARY AND CONCLUSIONS

Solid modeling is the most promising technology for defining mechanical components and products unambiguously if certain theoretical gaps (notably tolerancing) and technological limitations (geometric coverage, speed, complexity limits) can be overcome. Contemporary solid modeling systems provide good support for analytical procedures that can be used to verify final designs and to optimize parametric (nearly final) designs. However, current systems do not provide much support for the conceptual and preparametric phases of design, which are wholly unautomated at present. Human designers may find a future generation of systems that admit incompletely specified solids, "implied" solids, and solids defined through constraints to be more congenial, but difficult research problems must be solved before such systems appear.

Automation of the manufacture and assembly of mechanical goods is progressing systematically, with two kinds of modeling playing key roles. Solid modeling provides unambiguous definitions of what is to be made and also provides directly or through coupled analytical procedures models of the effects of processes on solids. Lower-level (feature, process) models provide primitives for planning automata—now wholly in the research stage—that eventually should produce complete sets of plans and programs for making, inspecting, and assembling parts automatically.

Now consider design: Mechanical-design automation and, more fundamentally, the understanding of mechanical design in a scientific sense are progressing slowly if at all. Thus, we have a growing technological imbalance, with manufacturing striding ahead of design in terms of both scientific understanding and automation. One of the major gaps in the understanding of design is the lack of means for modeling mechanical "function" in a manner that links function to form.

APPENDIX

NOTES ON THE EVOLUTION OF DESIGN AND MANUFACTURING

In the view of design and manufacturing developed above, form is central. It defines a part or product as a spatial entity and, when a material specification is added, as a physical entity. Form is induced from function by designers using processes we understand poorly. Fabrication is induced from form by manufacturing planners, using processes we understand better but still not well enough. Broadly speaking, the backward mappings—from fabrication to form through process simulation, and from form to function through analysis—are better understood than the forward mappings.

In current industrial practice, form specifications ("designs") carry no explicit representations of function and no explicit specifications for manufacturing and assembly. Thus, modern part prints and assembly drawings or their solid-modeling equivalents include no descriptions of what parts are supposed to do and how they interact functionally (as opposed to spatially) with other parts. Similarly, there are no form specifications such as "Mill Slot A 1 inch wide" or "Mill Slot A of Part B to mate with Slider C of Part D." In current practice, holes, slots, and almost all aspects of form are defined wholly geometrically through toleranced parameters of surface subsets. (Threads, knurls, and a few other "process-defined" features are exceptions.)

The current focus on pure form as the medium for design specification is a recent development. We shall review briefly the evolution of design and manufacturing to

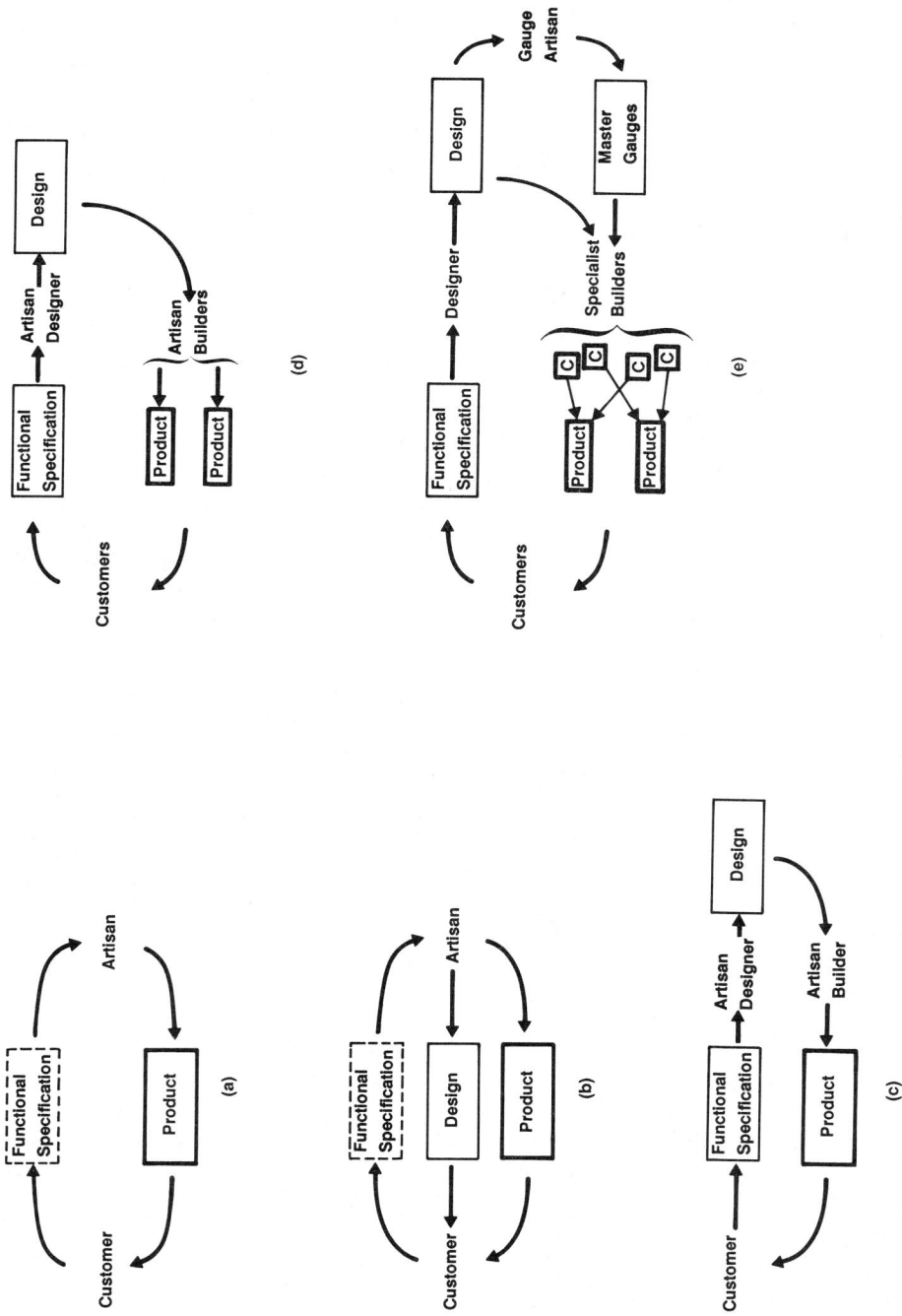

FIGURE 20 Evolution of the product cycle: (a) no explicit specifications; (b) explicit designs prepared by the artisan; (c) artisan designers become distinguishable from artisan-builders; (d) adaptive serial assembly; (e) artisans are replaced by specialist builder and assemblers.

see how we arrived at our current situation.[11]

In the beginning (see Figure 20a) a product was its design—i.e., there were no explicit, separable specifications that could be called "a design for a product"; further, the designer and builder (and sometimes the customer) were one and the same. But explicit designs soon appeared (Figure 20b), usually as crude physical models or sketches prepared by the artisan(s) before launching into construction. As products became more elaborate, a class of artisan-designers (for example, the master cathedral builders and master shipwrights) became distinguishable from the artisan-builders (Figure 20c), and customers' specifications took on a more formal structure that evolved toward contractual descriptions and performance specifications.

Through most of history, multiple copies of a product were almost always built by craft methods; i.e., single artisans or teams built a whole product from start to finish, as suggested in Figure 20d. This mode of fabrication, whether done in parallel, as in Figure 20d, or singly for one-off products, as in Figure 20c, may be termed adaptive serial assembly (colloquially, "file-and-fit") because (see Figure 21a) parts were manufactured to fit the evolving assembly. A major change came in the mid-nineteenth century, when the doctrine of gauge-based interchangeability began to be adopted. Figure 20e shows that versatile artisans were replaced with specialist builders of standard parts and specialist assemblers and that "gauge artisans" appeared as the precursors of manufacturing planners; products emerged through the intrinsically parallel, hierarchical tree of Figure 21b. From Figure 20e it is a relatively short step to the contemporary product cycle shown earlier in Figure 1.

---

[11]The history of technology is a young field that is only beginning to study the history of manufacturing in a systematic manner and has touched design only in a few isolated areas.

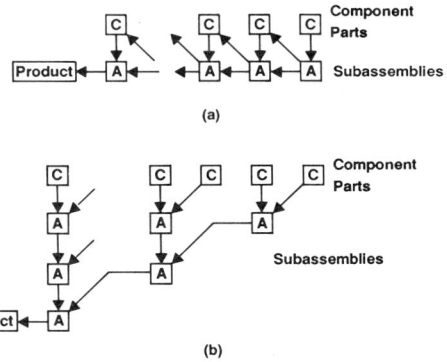

FIGURE 21 Serial and parallel assembly: (a) "file-and-fit" serial assembly; (b) hierarchical assembly of interchangeable parts.

Figure 22 shows how the component technologies in the product cycle evolved over time.

Manufacturing Technology

Because cutting was the dominant manufacturing process for most of history, the right-hand column in Figure 22 covers only cutting technology. (Casting is the other manufacturing technology with a long history.) The main trend is evident (Rolt, 1965, 1970; Woodbury, 1972): The free cutters of antiquity (early equivalents of a jackknife or single-edge cutting bit), with which persistent artisans could make almost anything, were gradually enveloped in machines that progressively reduced the need for human dexterity and strength. The change from animate to inanimate power was pure progress, but dexterity requirements were reduced at the cost of restricting a free-cutter's versatility through motional constraints imposed by mechanical guides. Much of the lost versatility was recovered, however, and precise repeatability was added, when separately guided motions were coupled mechanically, as in screw-cutting lathes, and later made electronically commandable, as in modern NC machines.

It is worth noting that the introduction

Time	Functional Representation and Design Synthesis Tools	Design Analysis Tools	Design Representation Technology	Manufacturing Planning Technology	Manufacturing Technology	Time
1000 BC			Crude physical models and sketches		Free cutters	1000 BC
					Guided cutters	
0 BC		Geometry	(Stereography)			0 BC
1000 AD			Ad hoc evolution			1000 AD
			Parallel projection			
1400 AD			Notions of perspective		Water-powered cutters	1400 AD
1600 AD					Coupled-motion cutters	1600 AD
		Trigonometry	Algebraic Geometry (Descartes, 1638)			
1700 AD				Mechanical programing (Jacquard, 1728)	Steam-powered cutters → Mechanization	1700 AD
1800 AD		Solutions of differential equations	Descriptive Geometry (Monge, 1795) Isometric Perspective (Farish, 1820)	Master gauges → Interchangeability	Precision metrology	1800 AD
		Numerical methods	Drafting conventions	Material science	Mechanically programed cutters	
1900 AD	Standardization of common components	Finite-difference	Drafting standards		Electrification	1900 AD
		Finite-element	Computer graphics Tolerancing 1929 (French) 1935 ASA Z14.1 1940 MMC 1944 True position	Group technology	Electronic control	
1950 AD		Computer simulation	Solid modeling 1973 ANSI Y14.5	NC and robot programing languages	NC machine tools Industrial robots	1950 AD

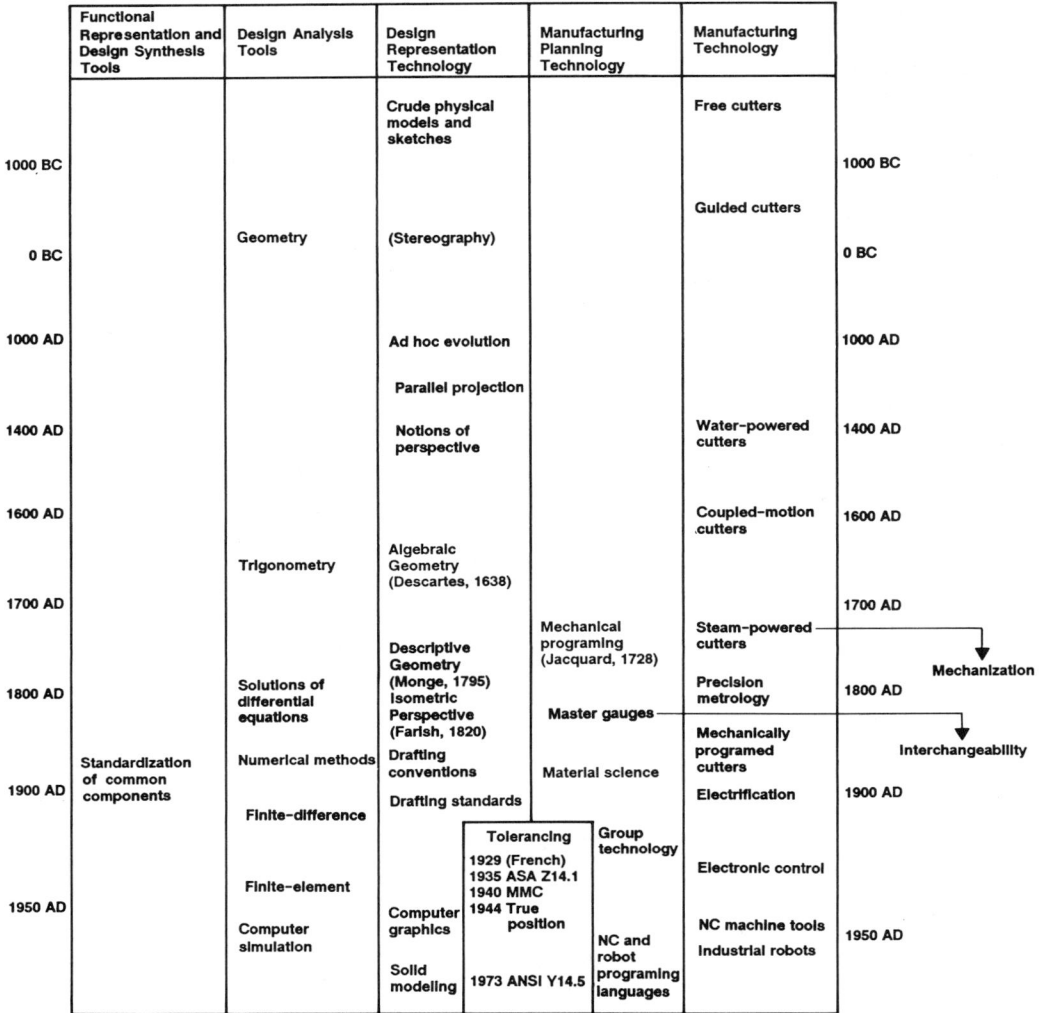

FIGURE 22   Historical evolution of component technologies in the product cycle.

of steam power liberated early factories from the tyranny of locale imposed by water power and thus sparked the wave of mechanization that is often called the (First) Industrial Revolution. (Factory electrification around 1900 fostered further decentralization by liberating machines from mechanical couplings to central factory steam engines.) Broadly, mechanization caused human dexterity and muscle to be replaced with mechanically induced precision and power, but humans functioning as planners, sensors, and controllers of machines continued to be essential components of manufacturing enterprises. The Second Industrial (or Automation) Revolution, whose advent may be marked by the invention of computers and NC machines around 1950, is now eliminating the need for human planning, sensing, and control.

Manufacturing Planning Technology

Codified history in the area of manufacturing planning technology is at best fragmentary, with Hounshell's recent book (1984) standing as an exemplary treatment of the evolution of interchangeability and Noble's polemic (1984) offering a biased but interesting view of the emergence of NC technology.

One might mark the advent of planning technology by the Jacquard loom's use of mechanical programming—a textile technology that took more than a century to propagate into the mechanical industries (e.g., in the form of screw machines). The most notable event was the introduction of master-gauge principles in American armories in the early 1800s; this led to interchangeability and "the American system of manufacture." The rise of metal-cutting science and material science in the later 1800s fostered engineering assignment of manufacturing parameters (e.g., speeds and feeds in machining) rather than cut-and-try values. Group technology appeared in Europe early in the current century and, after considerable elaboration in different contexts (design standardization, manufacturing-method standardization, etc.), remains controversial and largely devoid of scientific foundations. The invention of computers, NC machines, and industrial robots at midcentury led to the development of NC and robot programming languages, and some current research is focused on developing manufacturing-planning automata that can write NC and robot programs automatically.

Design Representation Technology

Crude models and sketches were used to represent man-made artifacts from the earliest eras of recorded history. In the late pre-Christian era the Greeks, through studies of astronomical stereography, had the tools to develop a systematic engineering graphics, but did not do so. Booker (1979) speculates on why they did not do so and provides a great deal of additional information in this area. Technical graphics, as we now know it, began to evolve in a largely ad hoc manner in the first half of the present millennium. Figure 23 shows typical twelfth-century "practice," and Leonardo's drawings exemplify the highest craft.

Descartes laid the foundations for modern geometry in the 1600s by coupling algebra and geometry, and Monge's "descriptive geometry" set in the late 1700s the main techniques and conventions used in modern engineering graphics. (Monge's work was regarded as so important that for a period in Napoleonic times it was classified as a military secret.) Figure 24 (Booker, 1979) shows an English drawing of 1804 that defines part of the valve gear for one of Richard Trevithick's steam engines. Observe that multiview parallel projection and section views were well established and that there are no dimensions on the half-size assembly drawing. There were no detailed drawings for the separate parts of the assembly, and the valve was made by a craftsman who scaled the assembly drawing using a proportional divider.

Drafting conventions—for example, American third-angle drawing-layout conventions—and dimensioning practice evolved in the later 1800s and began to be standardized in the 1900s. Drafting began to be "computerized" in the late 1950s with the advent of computer graphics, and now drafting is being replaced by solid modeling as the primary medium for design specification.

The evolution of tolerancing practice warrants special discussion. Levy (1974) notes that tolerancing was not introduced into American industry until about 1900 and that the first mention of the subject appeared in one paragraph in the 1929 edi-

FIGURE 23   A twelfth-century drawing of an undershot waterwheel.
Drawing is from "Hortus-Deliciarum," a manuscript containing drawings compiled by the Abbess Herrad of Landsperg in about 1160.

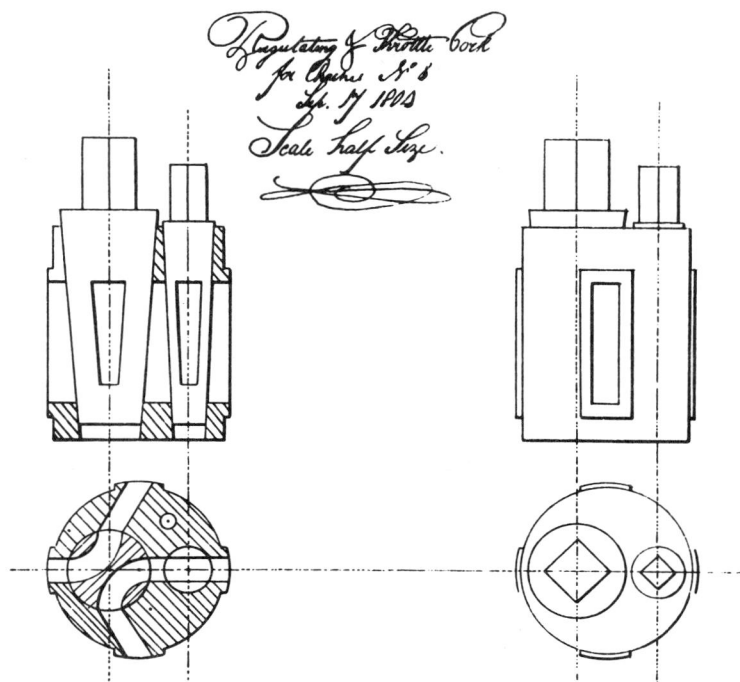

FIGURE 24   An undimensioned drawing of 1804. SOURCE: Science
Museum, London.

tion of French's classic textbook on engineering graphics, now in its tenth edition (French and Vierck, 1966).

The first American drafting standard, ASA Z14.1, appeared in 1935 as an 18-page document containing two paragraphs on limit ("plus-minus") tolerancing. The development of "geometric" or "true-position" tolerancing was done largely in Europe in the 1930s and 1940s. Chevrolet brought maximum material condition (MMC) concepts to this country in 1940, and Parker and Gladman (Levy, 1974) codified true-position principles in Britain in 1941–1944. The new system was refined in a series of draft standards that culminated in this country in ANSI Y14.5 in 1973, and

that standard has subsequently been amended in detail.

Figure 25 shows the rapid evolution of tolerancing practice. The 1920-vintage drawing of Figure 25a contains no explicit tolerances but specifies manufacturing processes (bore) for two holes and a precision for one hole through a functional requirement—running-fit (R.F.) on Spindle C. The 1950s drawing of the same part in Figure 25b specifies limit tolerances for the two holes and requires that a face be perpendicular RFS ("regardless of feature size"—Y's size, in this case) to hole Y. The 1970s drawing in Figure 25c retains the limit tolerances (but with subtly different interpretations) and the squareness toler-

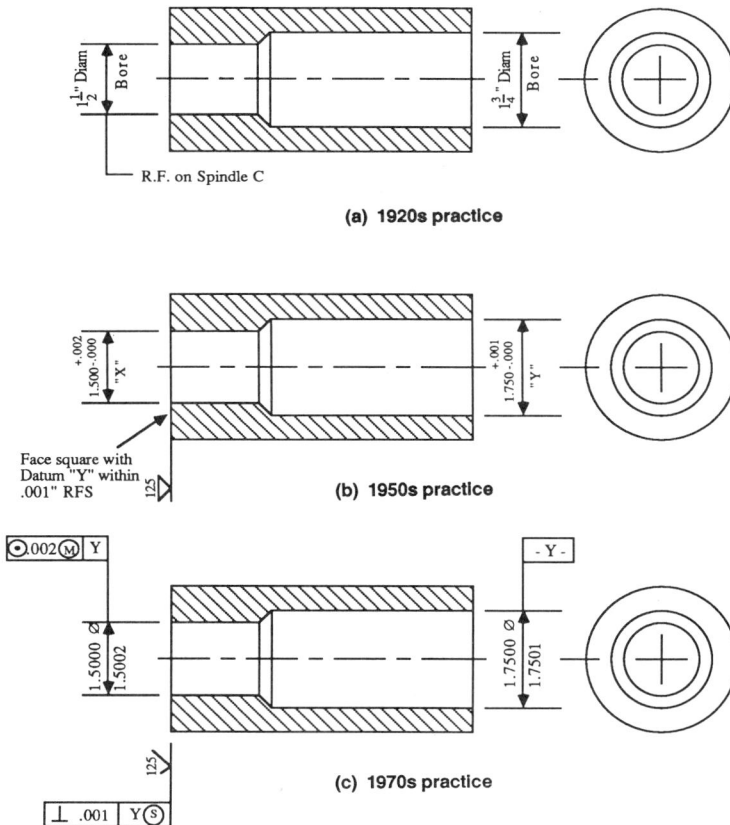

FIGURE 25   Evolution of tolerancing practice. Reproduced from *A History of Engineering Drawing* by P. J. Booker, courtesy of the author and Mechanical Engineering Publications.

ance, and also requires that the smaller hole be concentric with the larger to an MMC tolerance.

Figure 26 shows a table of symbols for interpreting Figure 25c, but more importantly, it illustrates the direction of modern tolerancing practice, which is to specify allowable "size," positional, and relational variations on or between "features" of parts. Alas, all of this work on tolerancing lacks mathematical foundations and thus, in the current era of informationally complete solid modeling, tolerancing stands as an open problem in engineering science.

Design Analysis Tools

The evolution of design analysis tools followed closely developments in mathematics, physical science, engineering science, and now computer science. Euclidean geometry became available in pre-Christian times, and trigonometry became a practical tool after Arabic numerology and arithmetic had been adopted. Analytical activity accelerated in the 1700s and thereafter when Euler, Fourier, and others developed methods to solve Newtonian differential equations in simple domains; this work laid the foundations for the development of numerical methods in the nineteenth and twentieth centuries. The invention of prac-

tical mechanical calculators about a century ago brought numerical methods into engineering use, and over the past 20 years electronic digital computation has brought finite-element and digital-simulation methods into widespread engineering use. Thus, the near-term future in design analysis seems to lie with increasingly powerful and parallel numerical computation. But limits set by the physics of known computing components, and also by the asymptotic complexity of known algorithms, are on the horizon, and we might see a return to quasi-analog methods through such routes as connectionist machines.

Functional Representation and Design Synthesis Tools

The first column in Figure 22 has but one, weak entry—the standardization of common components. The dilemma here is that engineers often can represent the "functionalism" of a product phenomenologically—through stress/strain equations in structures, heat-transfer and energy-exchange equations in combustors, and so forth—but we have almost no mathematical couplings from phenomenological models to the forms of artifacts that can exhibit specific phenomena. (Proceeding from form to function through analysis is easier, at least in principle, as we have already noted). Thus, the induction of physical form from mechanical function—mechanical design synthesis—continues to stand as an ill-understood "human creative activity."

COMMENTS

There are some striking trends, facts, and anomalies in the history just summarized. Five are considered here.

• Military needs have stimulated progress in design and manufacturing from the earliest to the most modern times. To cite but three examples: The first significant

	TYPE OF TOLERANCE	CHARACTERISTIC	SYMBOL
FOR INDIVIDUAL FEATURES	FORM	STRAIGHTNESS	—
		FLATNESS	▱
		CIRCULARITY (ROUNDNESS)	○
		CYLINDRICITY	⌀
FOR INDIVIDUAL OR RELATED FEATURES	PROFILE	PROFILE OF A LINE	⌒
		PROFILE OF A SURFACE	⌂
FOR RELATED FEATURES	ORIENTATION	ANGULARITY	∠
		PERPENDICULARITY	⊥
		PARALLELISM	//
	LOCATION	POSITION	⊕
		CONCENTRICITY	◎
	RUNOUT	CIRCULAR RUNOUT	↗ •
		TOTAL RUNOUT	↗↗ •

FIGURE 26   Attributes covered by modern tolerancing practice.

water-driven cutting machines were cannon-boring mills; the codification of drafting and tolerancing standards was driven largely by military procurement needs; the development and early dissemination of NC technology were undertaken to meet the stringent technical requirements posed by military supersonic flight.

• Mechanization and interchangeability are often thought to be linked or even synonymous practices, but they are in fact independent. Neither implies or requires the other.

• The century-long gap between the rise of interchangeability and the codification of tolerancing seems at first astonishing, for how can one manufacture interchangeable parts without strict manufacturing tolerances? The key lies in understanding gauging. Parts built to fit adequately designed gauges will be precise in the sense of being interchangeable, but they may not be "accurate" in the sense of meeting measurements to an absolute standard external to the set of gauges. One of the technical mysteries in manufacturing history is how early tooling engineers (the gauge artisans of Figure 20e) designed adequate gauges in the pretolerance era.

Today the relationship between gauges and tolerances is clear: Gauges are built to reflect tolerances given in design specifications, and hence tolerances implicitly specify inspection procedures.[12] Thus, we have, for example, "Principle 5: The gage designer should not have to make arbitrary decisions regarding gage element size or location, since a complete product specification dictates these design and interchangeability criteria. The drawing is not complete if such decisions are required . . ." (Roth, 1970, 5).

• The prohibition in design specifications of explicit links to either function (R.F. on Spindle C in Figure 25a) or manufacturing

---

[12]Whether they do so unambiguously is an open research question.

processes (Bore in Figure 25a) is largely a post-World War II development. It was promoted vigorously by military agencies to facilitate multisource competitive procurement and also as interchangeability insurance; it was subsequently adopted by many civil manufacturers for roughly the same reasons. The result is the current practice of defining parts and assemblies almost wholly through geometrical mechanisms. (Companies that have retained explicit links to function or process in their design specifications are almost always vertically integrated and do little outsourcing.) Those who advocate the currently fashionable doctrine of "design the process with the product" should be aware of the consequences of applying that doctrine too rigidly, because in some senses it is a retrograde step.

• We conclude these comments by noting again the lack of understanding of methods by which form can be induced from function. Architects have pondered this problem for several decades (Alexander, 1964; Habraken, 1987), and over the past decade mechanical engineers have attacked the problem on several fronts (Dixon et al., 1987; Rinderle 1986, in press; Suh et al., 1981; Ullman et al., in press).

REFERENCES

Alexander, C. 1964. Notes on the Synthesis of Form. Cambridge, Mass.: Harvard University Press.

Bezier, P. 1972. Numerical Control: Mathematics and Applications. New York: Wiley.

Booker, P. J. 1979. A History of Engineering Drawing. London: Northgate Publishing.

Brown, C. M. 1982. PADL-2: A technical summary. IEEE Computer Graphics and Applications 2(2):69–84.

Chan, S. C. 1987. MPL: A New Machining Process/Programming Language. Ph.D. dissertation, University of Rochester. Available as Technical Report CPA-1 from COMEPP, Kimball Hall, Cornell University.

Chan, S. C., and H. B. Voelcker. 1986. An introduction to MPL—A new machining process/programming language. Pp. 333–334 in Proceedings of the

1986 IEEE International Conference on Robotics and Automation. New York: Institute of Electrical and Electronics Engineers.

Coons, S. A. 1967. Surfaces for Computer Aided Design of Space Forms. Report MAC-TR-41, Project MAC. Cambridge, Mass.: Massachusetts Institute of Technology.

Dawson, P. R. 1986. Modeling plastic flow and microstructure development during forming processes. Pp. 157–160 in Proceedings of the Thirteenth NSF Grantees Conference on Production Research and Technology. Dearborn, Mich.: Society of Manufacturing Engineers.

Dixon, J. R., J. J. Cunningham, and M. K. Simmons. 1987. Research in designing with features. Proceedings of the IFIP Workshop on Intelligent CAD, Massachusetts Institute of Technology, October 1987.

French, T. E., and C. J. Vierck. 1966. A Manual of Engineering Drawing for Students and Draftsmen, 10th ed. New York: McGraw-Hill.

Goldfeather, J., and H. Fuchs. 1986. Quadratic surface rendering on a logic-enhanced frame-buffer memory. IEEE Computer Graphics and Applications 6(1):48–59.

Habraken, N. J. 1987. Concept design games. A Workshop: The Study of the Design Process, M. B. Waldron, ed. Columbus: Ohio State University. In press.

Hartquist, E. E., and H. A. Marisa. 1985. PADL-2 Users Manual. UM-10/2.1, Cornell Programmable Automation. Ithaca, N.Y.: Cornell University.

Hounshell, D. A. 1984. From the American System to Mass Production, 1800–1932. Baltimore: Johns Hopkins University Press.

Kedem, G., and J. L. Ellis. 1984. Computer systems for curve-solid classification and solid modelling. U.S. Patent Application.

Kela, A. 1987. Automatic finite element mesh generation and self-adaptive incremental analysis through solid modeling. Ph.D dissertation, University of Rochester.

Kela, A., R. L. Perucchio, and H. B. Voelcker. 1986. Toward automatic finite element analysis. ASME Computers in Mechanical Engineering, 5(1):57–71.

Levy, S. J. 1974. Applied Geometric Tolerancing. Beverly, Mass.: TAD Products Corporation.

Noble, D. F. 1984. Forces of Production: A Social History of Industrial Automation. New York: Alfred Knopf.

Requicha, A. A. G. 1977. Part and Assembly Description Languages: I—Dimensioning and Tolerancing. Technical Memorandum 19, Production Automation Project, College of Engineering and Applied Science, University of Rochester.

Requicha, A. A. G. 1984. Representation of tolerances in solid modeling: Issues and alternative approaches. Pp. 3–19 in Solid Modeling by Computers—From Theory to Applications, M. S. Pickett and J. W. Boyse, eds. New York: Plenum Press.

Requicha, A. A. G., and H. B. Voelcker. 1982. Solid modeling: A historical summary and contemporary assessment. IEEE Computer Graphics and Applications 2(2):9–24.

Requicha, A. A. G., and H. B. Voelcker. 1983. Solid modeling: Current status and research directions. IEEE Computer Graphics and Applications 3(7):25–37.

Rinderle, J. R. 1986. Implications of form-function-fabrication relations on design decomposition strategies. Pp. 193–198 in Proceedings of the 1986 Computers in Engineering Conference, Vol. 1. New York: American Society of Mechanical Engineers.

Rinderle, J. R. 1987. Function, form, fabrication relations in design. In A Workshop: The Study of the Design Process, M. B. Waldron, ed. Columbus: Ohio State University. In press.

Rolt, L. T. C. 1965. A Short History of Machine Tools. Cambridge, Mass.: MIT Press.

Rolt, L. T. C. 1970. Victorian Engineering. Middlesex, England: Penguin Books.

Roth, E. S. 1970. Functional Gaging, 2d ed. Dearborn, Mich.: Society of Manufacturing Engineers.

Samuel, N. M., A. A. G. Requicha, and S. A. Elkind. 1976. Methodology and Results of an Industrial Part Survey. TM-21. Rochester, N.Y.: Production Automation Project, University of Rochester.

Suh, N. P., J. R. Rinderle, H. Nakazawa, and K. Kaneshige. 1981. An axiomatic approach for improving design and manufacturing processes. Pp. 0-1 to 0-8 in Proceedings of the Ninth NSF Grantees Conference on Production Research and Technology. Dearborn, Mich.: Society of Manufacturing Engineers.

Sungurtekin, U., and H. B. Voelcker. 1986. Graphical simulation and automatic verification of NC machining programs. Pp. 156–165 in Proceedings of the 1986 IEEE International Conference on Robotics and Automation. New York: Institute of Electrical and Electronics Engineers.

Ullman, D. G., L. A. Stauffer, and T. G. Dietterich. 1987. Preliminary results of an experimental study of the mechanical design process. In A Workshop: The Study of the Design Process, M. B. Waldron, ed. Columbus: Ohio State University Press. In press.

Voelcker, H., A. Requicha, E. Hartquist, W. Fisher, J. Metzger, R. Tilove, N. Birrell, W. Hunt, G. Armstrong, T. Check, R. Moote, and J. McSweeney. 1978. The PADL-1.0/2 system for defining and displaying solid objects. ACM Computer Graphics 12(3):257–263.

Wang, K. K., S. F. Shen, C. A. Hieber, R. C. Ricketson, V. W. Wang, S. Emerman, and C. Cohen. 1986. Integration of CAD/CAM for injection-molded parts. Pp. 93–97 in Proceedings of the Thirteenth NSF Grantees Conference on Production Research and Technology. Dearborn, Mich.: Society of Manufacturing Engineers.

Wolfe, R. N., M. A. Wesley, J. C. Kyle, F. Gracer, and W. J. Fitzgerald. 1987. Solid modeling for production design. IBM Journal of Research and Development 31(3):277–294.

Woodbury, R. S. 1972. Studies in the History of Machine Tools. Cambridge, Mass.: MIT Press.

THE STRATEGIC APPROACH TO PRODUCT DESIGN

Daniel E. Whitney, James L. Nevins, Thomas L. De Fazio,
Richard E. Gustavson, Richard W. Metzinger,
Jonathan M. Rourke, and Donald S. Seltzer

ABSTRACT   The strategic approach to product design (SAPD) is a multistep process that seeks to implement integrated product and process design. Because of the inherently integrative nature of the assembly process, that is the focus of SAPD. This paper outlines the steps of SAPD, compares it to conventional product design methods, and suggests research that is needed to provide analytical and computer support to what is at present a team approach dependent on experts.

INTRODUCTION

There is growing concern that U.S. manufacturing is no longer competitive with that of other countries in the global marketplace (National Research Council, 1986). Many causes for this concern have been presented (Olmer, 1985; Porter, 1986; Wheelwright and Hayes, 1985), and many remedies have been proposed. The important point is that, "despite the enthusiastic claims of technology developers and vendors, technology alone will not improve competitiveness" (National Research Council, 1986, ix). What is needed is an integrated approach to manufacturing systems supported by new technologies. This paper discusses a strategic approach being used by some U.S. and many Japanese companies to raise their productivity by both integrating their design and manufacturing functions and modifying their manufacturing institutions.

A new generation of advanced manufacturing technology, most notably flexible manufacturing systems (FMSs), is having an important impact on unit operations. This is particularly true of product design that is carried out through CAD systems that are part of an FMS. Although such systems allow many operations to be carried out automatically, the impact of these new approaches on the functional way in which most companies operate has been minimal.

These conditions can be contrasted with what is occurring when companies consider the next generation of assembly, namely, automated flexible assembly. There is a growing awareness that automation for assembly cannot be treated in an isolated manner. Assembly, with its close coupling to design, vendor control, quality control, etc., requires a new, more highly integrated approach to manufacturing. Assembly is inherently integrative. Parts that were separately designed, made, handled, and inspected must be joined together, handled together, and tested together. The assembly process focuses attention on pairs and groups of parts. Decisions that affect assembly also affect nearly every other aspect of production and use of a product. Assembly, therefore, is a natural forum for launching an integrated attack on all the phases of a product, from conception and fabrication to quality and life cycle. The recent interest in design for assembly is an outgrowth of the realization that assembly is an important phase in the life cycle of a product. Other phases, however, are also important. They may last longer and cost more. Thus, design for assembly cannot be done in isolation.

An approach that integrates product design and all aspects of the manufacturing process can only be accomplished through a strategic approach to product design (SAPD). Such an approach allows the entire system to be rationalized. Furthermore, it helps identify the need for CAE data bases and computerized design tools that are needed to support this tightly integrated activity.

SAPD provides the opportunity to deal with the many trade-offs that must be made. It also allows it to be done at the best time, namely, when the product is being designed. Although design for assembly usually deals with single parts, SAPD deals with groups of parts, subassemblies, and the total product.

SAPD is independent of any particular assembly technology. It deals with many nonassembly issues, and it provides rationally designed products. This rationality benefits assembly by any technique, including manual; it benefits other phases of production, such as inspection; and it

can materially affect the life cycle of the product.

Although many companies recognize the advantages of an integrated approach to product design, Japanese firms have been the most proficient in using it, some for 20 or 30 years. A number of advanced U.S. companies have recently embraced this approach.

The remainder of this paper provides two views of SAPD. The intuitive approach, as it is usually practiced in both the United States and Japan, is described first. Second, the intellectual ingredients of this approach are extracted and listed so that the research problems can be seen more clearly. Finally, some of the educational, curricular, and technology transfer implications are discussed.

Before the process is described in detail, it is important to give some of the context.

DRIVERS OF CHANGE IN MANUFACTURING

Part fabrication is essentially a series of independent steps with minimum interconnections between operations being performed on individual piece parts. These operations are intended to enforce a particular geometric configuration on formable materials. The ideal geometry exists in the design. Because of statistical variations in both the process and the materials, the fabrication process creates a part whose properties only approximate those of the ideal. These perturbations must be controlled so that the resulting part has properties that fall within design tolerances.

Models for fabrication processes are generally good. Thus, the degree to which uncertainty perturbs the ideal geometry can be predicted fairly well—certainly well enough to allow automatic fabrication to proceed with high confidence. The level of understanding of assembly and inspection, however, is much less mature.

The technology of part fabrication has advanced considerably in the past 30 years. In the early days, single numerically controlled machines took their instructions from hand-delivered paper tapes. Today, systems of 10 or 20 such machines work together to make groups of parts that were designed on computers. The tapes have been replaced by direct data links. The designs are now supported by three-dimensional geometric modeling programs and by computers that hold information on such things as materials, stress analysis, and machining methods. Engineers and researchers currently face a problem of logistics and scheduling—that is, finding the right mix of parts to keep the machines busy. Few tools are currently available to aid in solving these problems.

The production capacity of such automated systems is impressive, and so is the speed with which parts can be designed and made. Based on the performance of these design and fabrication systems, most manufacturers are convinced of the benefits of close integration between design and fabrication. However, too few manufacturers recognize that design for fabrication is not the same as design for overall producibility. Design for fabrication, sometimes mistakenly called design for manufacturability, considers parts as isolated entities rather than in groups that must function together. SAPD seeks to correct that shortcoming.

Beyond this, several deeper problems have become visible. The logistical problems mentioned earlier have sensitized manufacturers to two needs: to integrate scheduling and system operation into their strategies, and to design parts and families of parts to make better use of these systems. Existing systems are not adequately flexible. They are suited to only one kind of manufacture, that of serial metal removal or metal bending on a limited variety and on a small number of carefully chosen pieces. Other kinds of manufacture, such as casting, composite material lay-up, and powder metallurgy, and other steps besides fabrication, such as assembly, have not yet been brought under the umbrella of computer-integrated systems. Extensions like these require much more careful product design and process organization.

The response to these challenges has been a surge of interest in DFA. Ten or twelve years ago the "correct" attitude of automation engineers was to take the product as it was designed and do their best to assemble it, either manually or by machine. Manufacturers tended to emphasize low part cost and fast assembly. The result was that all the problems introduced by fabrication or logistics, including parts out of tolerance, or late, or damaged, had to be solved by the ingenuity of the assemblers. This was the social structure of manufacturing—a sort of hidden agenda.

This hidden agenda made the introduction of robots and other advanced assembly technology either difficult or impossible. Manufacturers were reluctant to pay for the cost of extra equipment to support the robots, such as part feeders, palletized parts, and control computers. They were also disappointed that robots were not as resourceful as people in handling problems such as out-of-tolerance parts. Few companies wanted to spend the extra money for higher quality parts just so robots could put them together. Instead they demanded better and more intelligent, adaptive robots that could solve these problems.

In the past few years these attitudes have begun to change. Manufacturers are realizing that, although better quality parts make possible assembly by robots, they also create a better quality product. They are realizing that using the assembly system as a filter to detect bad parts is a bad way to run a factory. The hidden agenda is gradually being stripped away. With this is coming a deeper understanding of the role of technology in products and processes. In addition, we see the convergence of forces that will demand new approaches. These forces are

• The complexity of new products and the disappearance of the learning curve;
• The complexity of modern worldwide production and the changing nature of competition; and
• The disappearance of manual assembly as an option.

Complexity of Products and the Disappearance of the Learning Curve

Modern products can contain thousands of parts and many technologies. A new automobile can take more than 5 years from initial specifications to production. A new surface combat ship—probably the most complex item built today—can take up to 10 years to cover the same process. Modern products are characterized by combinations of energy and information storage systems. They may be made of materials that are not merely transformed from mined ores or feedstocks but are created with properties that especially serve the needs of the product. Some new products are tailored to specific market niches, thus demanding small production volumes. Thus, there is a need for rapid advances in a product line, fast updating of designs, and quick changes in production schedules. The learning curve that allowed design or production problems to be worked out over a long period of time has been compressed or eliminated. Instead, learning must be spread over a series of products. To be valuable, this learning must focus on generic issues rather than product-specific issues so that the lessons can be passed on to the next product. Two responses to the disappearance of the learning curve can be identified.

First, manufacturers have responded to the time compression by doing things faster. Products and systems are being de-signed more rapidly, in part through better planning and more effective computer tools. Better planning means organizing the design process so that more factors are taken into account early, reducing the chance of damaging surprises later. It also involves identifying a good sequence in which to make design decisions, thus retaining some room to maneuver in the later design stages and also permitting decisions to be made in a way that minimizes the need to iterate (Akagi et al., 1984). Better and more comprehensive computer tools, computer-aided-design, computer-aided engineering, and computer-aided manufacturing allow calculations to be made more rapidly and to cover more cases.

Second, manufacturers must think more deeply about what they do and the most effective use of the lessons learned from previous design activities. Although the learning may still take longer than the time available to design a new product or system, learning can occur on a continuous basis. The basis for the systemization is the recognition of generic or repeating elements in the product, the processes, or the design steps themselves.

An analogous approach, called group technology (Opitz, 1967), may be found within the narrowly defined topic of machined parts. Although the application of the approach differs from case to case, the spirit of it is to recognize major similarities or differences between nonidentical items so that they can be grouped. Within a group, the similarities are used to advantage by means of sharing, for example, a machine or a measuring method. When facilities can be shared by a large number of items, time and cost are usually saved. Had the similarities not been recognized and the groups not formed, the facilities would have dealt with the items piecemeal, and no quantity savings could have been made.

To apply this idea in design, one needs a systematic, step-by-step approach that asks the same kinds of questions and requires repeated applications of the same kinds of analyses. As these design steps are repeated, one should become better at recognizing the similarities, even though each product or system is outwardly different. As the steps become clearer, it should be possible to develop computer tools for carrying them out. A major aim of this paper is to highlight these steps and identify both the existing tools and the gaps where more knowledge is needed. The formulation of these steps and their organization into a coherent approach constitute the intellectual challenge of manufacturing.

Complexity of Processes and the Changing Nature of Competition

Production processes are complex. Since many products are changing frequently, timing is critical. The result is that processes require more care and attention, and more data are needed to determine how a manufacturing system is performing. New production technology requires new skills and attitudes from workers and managers.

Product and process complexity arise from the appearance of new kinds of products, many of which contain a true mix or integration of mechanical and electronic functions, thus requiring more broadly educated product and process designers. An example of such a product is computer disk drives, whose success depends on careful design, precise tolerances, extreme cleanliness, fine timing and balance, and the skill and attention of dedicated people.

Successful competition for markets for such products demands a new response. In older industries or in those that have reached maturity, the basis of competition is usually production efficiency. The managers in such industries focus on asset management and make most of their decisions based on incremental economic criteria. In

newer industries, or in older ones faced with new competitors, the bases of competition are more likely to be product innovation, advanced technology, and quality. Although mature products probably differ little in technology and are distinguished by price, newer products may command a price premium based on quality or novelty. Manufacturing decisions in such industries are therefore less likely to be dominated by incremental economics and more by the ability of the product and factory to support the competitive strategy behind the product, such as the ability to evolve rapidly or be responsive to a changing market while maintaining a high product quality.

Disappearance of the Manual Assembly Option

Since manual metal removal was never an option, machinery for this purpose developed early in the industrial revolution. Thus, metal removal processes were among the first to be well understood. While manual assembly has been a common past practice, it is rapidly disappearing as an option in high-technology products. People have too much difficulty providing the required quality, uniformity, care, documentation, and cleanliness that are required of many of today's products. This is not to say that remarkable human performance is impossible. In Japan, for example, it is usual for a worker to make only one assembly error (wrong, missing, or broken part) in 25,000 to 100,000 operations. Since this is considered not good enough, the Japanese use these error numbers to justify further automation. Such is their conviction that modern products demand even higher quality.

Direct substitution of robots for people will not solve the problem, however. The successful use of robots requires a carefully designed environment consisting of a properly designed product, well-trained operators, and well-scheduled operations. Design for assembly is also not enough by itself. Since many parts or subassemblies will be purchased, the same kinds of problems must be solved by outside vendors.

The decision to replace a person with a piece of machinery, such as a robot, always involves some form of economic analysis. The commonly used analytic techniques assume that the substitution results in a system that is equivalent and interchangeable with the current system in every way except cost. Other factors need to be included, however. Since each proposed replacement is different in its ability to deliver quality, it is important that this be reflected in the analysis. Failure rates, repair costs, and testing strategies, for example, must be considered. These in turn are affected by product design, as discussed later. In high-technology products, the cost of materials is frequently more important than time or labor, so the ability of an assembly-test system to deliver a good yield is crucial to maintaining production volume and profit. Thus, economically justifiable manufacturing systems can contain both machines and people. The techniques of economic analysis must also be improved to the point that a proper assessment can be given of the true value of fully integrated assembly systems and new product designs. Economic measures, such as return on investment and machine replacement, must be supplanted by more sophisticated criteria.

All this means that we have to understand assembly as thoroughly as we now understand metal cutting. Indeed, all the processes in manufacturing—material handling, stocking, transport, inspection, judgment of suitability, and granting of "exceptions"—that are now routinely handled by people in an intuitive, judgmental, and often undocumented way, will have to be brought to a higher level of understanding, even if they are not to be executed by machines.

THE STRATEGIC APPROACH TO PRODUCT DESIGN

The way a product traverses its life cycle, including fabrication, purchase, assembly, inspection, use, repair, modernization, and disposal, is established when the product is designed. The effectiveness, efficiency, and cost of these various stages are all affected by decisions made during the design, which includes both product and process design. Since the product design must recognize strategic issues related to both the manufacturer and the user of the product, the required response has been called the strategic approach to product design. The remainder of this paper focuses on SAPD as a process and as a discipline.

The Educational Problem

Engineering schools teach a fairly straightforward version of how something is designed. This version of the design process is shown schematically in Figure 1. Engineers are given a technically oriented view that begins with the need for the product; proceeds to the preparation of product specifications, the making of trial designs, prototypes, and final designs; and concludes with a manufacturing process plan. There is a good deal of feedback as problems are uncovered and resolved. But in the main the process is self-contained from need to final design, with little outside interference.

This method suffers from too much linearity of the process, it is often too technical, and it is too compartmentalized, encouraging design to be the domain of the designer, manufacturing the domain of the manufacturing engineer, purchasing the domain of the purchasing manager, and so on.

A greatly improved method, one used by companies that are more successful in creating competitive products, shares at-

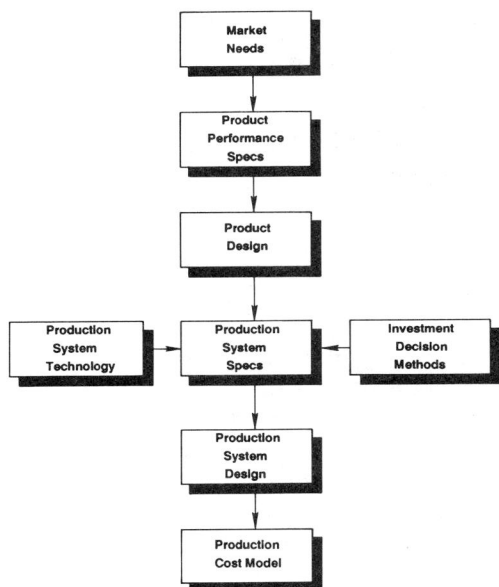

FIGURE 1 The conventional product design-production system design process.

tributes with the procedure shown in Figure 2. This method emphasizes the degree to which decisions made by the different parties affect other activities and alter the product's character (National Research Council, 1986, 102–112).

An example will illustrate the type of problems that may arise when one unit is unaware of the needs imposed by other units on the design of a product. Consider a particular military product that depends critically on an infrared detector. The purchasing department switched to a lower-price vendor without determining the reproducibility of the detectors that would be provided. Although subtle differences between detectors can significantly affect performance, these cannot be found until the product is partially assembled with optics, power supplies, and so on. To increase ruggedness and reduce cost, the unit is glued together, making disassembly to replace the detectors very expensive. Naturally, the product could be redesigned with threaded

joints to facilitate detector replacement as well as field repair. But the product in question is a single-use weapon. It must work the first time, its operating lifetime is only a few seconds, and its shelf life must be several years. Repair is simply not "in character" for this kind of an item. Thus, a decision to reduce costs by substituting one component for another can have consequences that can be appreciated only if the life history of the product is completely understood.

The point of this example is that a seemingly minor decision, made to optimize a corner of a company's operations, can have a pervasive effect on how a product is made or how it performs in the field. These decisions can completely defeat the designer's intentions. Management, engineering, purchasing, personnel, and manufacturing can each contribute to making or defeating a product.

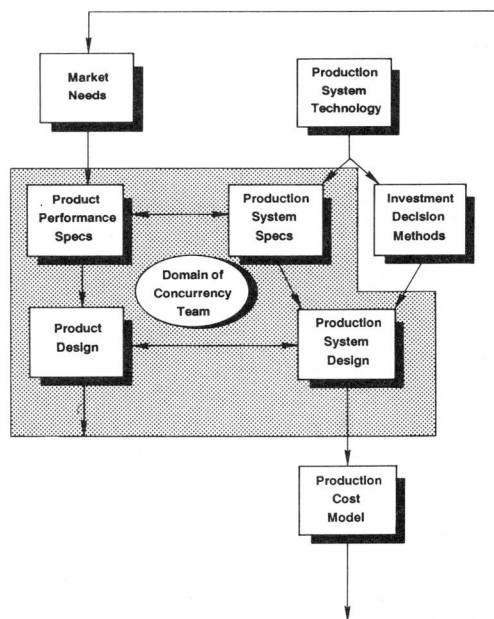

FIGURE 2 The emerging concurrency method of designing products and production systems.

Levels of Product Design Strategies

Product design can be divided into levels of activities that include functional design, manufacturing, and life-cycle considerations. Product designers traditionally do functional design. They choose materials, dimensions, and tolerances in such a way that the item will accomplish its intended purpose. In traditional organizations, the functional design is given to the manufacturing engineers to determine the processes for fabricating each part, including choice of machines, methods for maintaining tolerances, and make-or-buy decisions. The latter decisions essentially export to vendors some of the manufacturing engineers' problems. For this export to be successful, someone (e.g., the engineers or the purchasing agents) must carefully monitor the vendors. Irrational products and production systems can result if the monitors do not sufficiently understand the product and its requirements.

In addition, the manufacturing engineers must design the assembly system or method for the product. Traditionally, this is done by straightforward economic analyses, in which a choice is made between manual and "automatic" assembly. The latter is chosen usually in cases where the product is small, has less than 12 parts or so, and is made in quantities of about a million per year for several years.

In recent years, the need for improved competitiveness and productivity has led many companies to modify this process to recognize the entire life cycle of the product. When viewed in this manner, the process comprises life cycle issues such as product use, repair, and upgrading; manufacturing issues such as the assembly system, assembly operations, tolerances, and vendor control; fabrication issues such as make or buy and method of fabrication; and functional design issues. One consequence of this new view is that manufacturing engineers are involved earlier in the design.

Unfortunately, neither the more traditional nor the more recent sequence adequately describes practice. Converting a concept into a product is a complex procedure of many steps. As the design evolves, choices must be made concerning such things as materials, fasteners, coatings, adhesives, and electronic adjustments. Not only are these choices interdependent, it is likely that some of them would have been different if slightly different criteria for choice had been used by the designers. Furthermore, it becomes increasingly difficult as design proceeds to introduce new viewpoints and criteria without seriously delaying the design. Thus, if manufacturing engineers participate in the design debate from the start, their criteria can be properly considered. Similarly, if repair engineers, purchasing agents, and other knowledgeable people are represented, a better, more integrated design will result. In each case, the design will represent an interconnected web of decisions, and the participation of more parties will ensure that the web is better balanced.

To bring some structure to a detailed description of strategic design, the following breakdown of topics is used:

• The character of the product is determined; this identifies the attributes of the product.
• A study of the design for producibility and usability is carried out.
• A product function analysis is made to determine if the product's producibility and usability can be improved without impairing desirable functions.
• An assembly process is designed that includes a suitable assembly sequence, the identification of subassemblies, the integration of a quality control strategy with assembly, and the design of each part so that its functional tolerances and tooling toler-

ances (gripping and jigging surfaces) are compatible with the assembly method and sequence.

• A factory system is designed that fully involves the production workers in the production strategy, operates on minimum inventory, and is integrated with the methods and capabilities of the vendors.

Part fabrication, although essential, is not discussed in detail in this paper because it is a much more mature process.

Given the present state of design methodology and techniques, the best way to pursue these activities is to form teams of product designers and manufacturing engineers, with active participation by representatives from marketing, finance, purchasing, and personnel. Some companies call this process top-down analysis; others call it concurrency. To be successful, a concurrency team should be formed early and maintained in position until the product is test-marketed.

Figure 3 sketches the activities of the con-

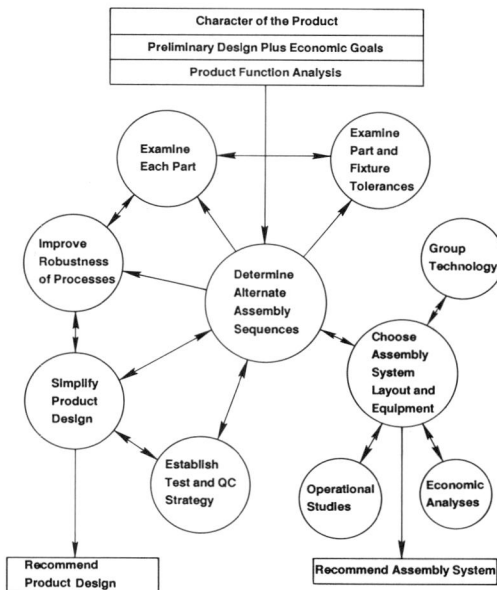

FIGURE 3 Outline of the strategic approach to product design, with emphasis on assembly issues.

currency team (omitting fabrication issues) and shows how the activities interact. The next few sections of the paper expand on some of these activities.

Character of the Product

A product's character is the combination of the basic features of how the product will be made, sold, and used. It must be determined early and recognized by everyone involved in the process. There is an endless list of possible features contributing to the character of a product. The following is an example of two product characters, together with their consequences for product design and production:

CHARACTER OF PRODUCT
1. Complex item, no model mix, used by untrained people, must have 100 percent reliability, used only once and thrown away.
    CONSEQUENCE
    Make high-quality parts, glue them together, do not try to fix after manufacture.
2. Complex item, contains a mixture of models and options, used by untrained people, lasts for years, and is serviced in the field.
    CONSEQUENCE
    Make high-quality parts, screw them together, and provide replacement parts and field repair service.

Design for Producibility and Usability

Once the character of the product has been defined, at least provisionally, true product design can begin. Design for producibility and usability is a top-down process. It is guided by the product, and it helps formulate the manufacturing strategy. This contrasts with many so-called examples of design for assembly, which are in fact just good (sometimes very good) re-engineering of the product itself without regard to an overall strategy. Innovative engineers can always come up with "improvements." Without a guiding strategy, there is no way to tell which improvements really support the strategy and which merely look like isolated improvements.

The main targets of the concurrency team are to

• Convert the product concept into a manufacturable, saleable, usable product design.
• Anticipate fabrication and assembly methods and problems.
• Simplify the design, fabrication, use, and repair by, for example, reducing the number of parts or identifying and increasing the number of parts common to different models.
• Improve the robustness of product and process by, for example, breaking product and process into self-contained modules, adjusting tolerances to eliminate chance failures, and identifying places where tests can be made.

It is readily apparent that design for producibility is different from value engineering, an activity aimed chiefly at reducing manufacturing cost by astute choices of materials or methods of making parts. Value engineering occurs after major product design is finished, and thus it is neither concurrent nor likely to be very thorough. The required thoroughness cannot be accomplished except through a concurrent process. More importantly, design for producibility includes and sometimes subordinates reduction of manufacturing costs within the larger goal of optimizing the entire life cycle of the product. The reason is that, although employees assemble it, customers or repair personnel may disassemble it, and the actions of these others can be made easier, safer, or more congenial to the character of the product by decisions made simultaneously with design or assembly decisions.

Design for producibility is also different from design for assembly. This activity, like value engineering, usually begins after the product is designed. It considers the parts one by one, simplifies them, combines some to reduce part count, or adds features to make fabrication or assembly easier. This can be characterized as a bottom-up process. It is guided by the parts rather than by a holistic concept of how the product is to be made and used.

Since the essence of a sophisticated design can depend on the careful choice of tolerances, materials, or novel fabrication methods that cannot be separated from the design of the manufacturing process, the concurrency process must begin early. Indeed, in some cases, the process is the design. An example of this process comes from Japanese shipbuilders. Their philosophy is that "design is a subset of production."

Shipbuilding is an intensely complex and time-consuming process (Chirillo, 1982; Whitney et al., 1986). The efficiency of shipbuilding is so heavily influenced by planning and organization that the Japanese have evolved a method that makes actual design of the ship a part of the construction planning process. Since welded joints are just as strong as the surrounding metal, the shape of the pieces that are welded and the location of the joints can be chosen to facilitate producibility, providing a new freedom in design. The Japanese carefully choose subassembly and module shapes to exploit efficient group-technology methods for making them.

Once the overall shape and characteristics of the ship have been determined, the size and shape of the pieces are determined by the order, method, or location in which they will be made. The size and shape of subassemblies into which those pieces will be built are first decided. Then the planners identify and give shape to the individual pieces of hull plate, pipe, deck, and so on, including the precise schedule of ordering raw material, joining of parts, and measuring to ensure that the assemblies will fit together the first time. Each of these subassemblies is called a zone, and all management, scheduling, cost accounting, and supervision is done by tracking these zones through several predetermined stages of production. Zones at a particular stage are grouped into similar areas, where each area constitutes a type of work with similar needs for human skills, machinery, measuring equipment, and so on. The ship is designed so that the maximum possible number of zones constitute areas that are easy to make.

Some examples, shown in Figure 4, indicate the levels of planning and production that the Japanese have introduced in parallel with their new designs. The relationship of the builders to the steel mills allows them to order the precise shape of plates they need, on short notice, and with the necessary uniform quality that permits carefully developed low-distortion welding methods to be used.

Many ships were built during the time that this procedure was being developed. Although ships differ in detail, the process not only emphasizes and takes advantage of the similarities but also encourages the designers to use these types of similarities and to identify rational ways of improving the producibility of later ships.

The full potential of design for producibility cannot be realized until concurrency team members fully understand how the product is supposed to work and be used. They achieve this understanding through analysis of product function.

Product Function Analysis

Product function analysis is an activity in which designers and engineers seek ways of simplifying or rationalizing a product's design by starting from what the product *should* do rather than how it performs that function *now* or how that function was performed in previous designs. Decisions regarding fabrication or assembly method can affect users as well as factory personnel and field costs as well as factory costs. These

decisions are related to design, not manufacturing. Because they affect the character of the product, they are strategic in their impact. No one department should make these decisions alone, nor can the decisions be parceled out for decentralized action. These are decisions that the concurrency team must address.

Assembly Processes

The concurrency team must also address assembly processes. Activities with strategic implications include

• Division of the product into subassemblies;
• Establishment of an assembly sequence;
• Selection of an assembly method for each step; and
• Integration of a quality-control strategy.

There is no set order in which to consider these activities, since the choices interact, and making them may trigger more design changes. It is convenient to discuss the first two of these activities together.

The choice of assembly sequence and the identification of subassemblies focuses attention on so many aspects of product design that they provide a natural starting point for integrative detailed design. Assembly sequence studies require identification of potential jigging and gripping surfaces, grip and assembly forces, clearances and tolerances, and other issues that must be accounted for in component design. Although these issues were not considered important when manual assembly was used, they are very important to machine assembly. For example, tolerances on grip and jig surfaces must be adequate with respect to mating surfaces. Tolerance adequacy can be determined using the Part Mating Theory that has emerged in recent years (CSDL Reports 1974–1980, R-800, R-850, R-921, R-996, R-1111, R-1276; 1979–1982, R-1407,

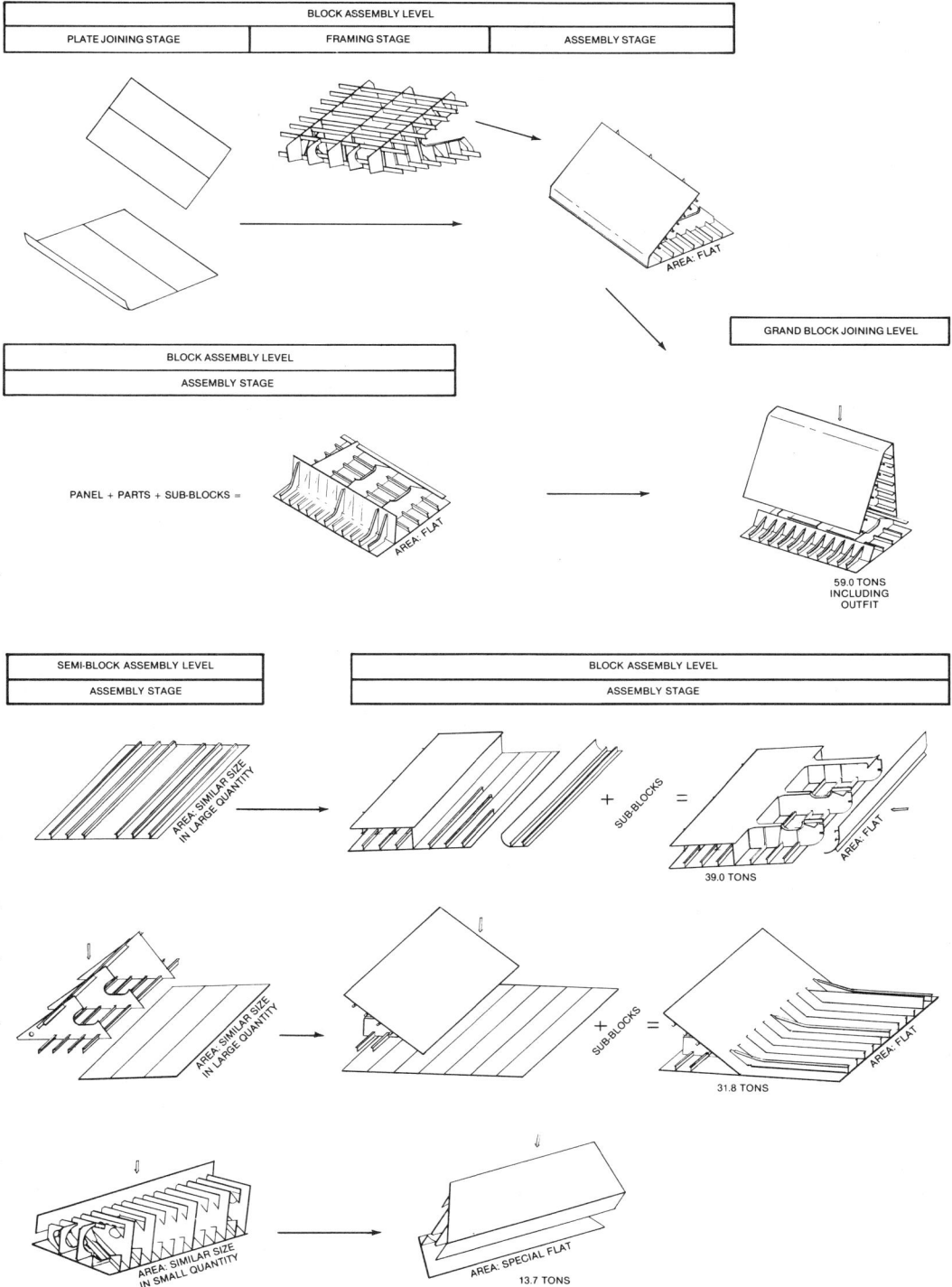

FIGURE 4   Example of division of ship hull into manufacturable subassemblies that may be grouped into like production classes. SOURCE: Chirillo (1982).

R-1537; Whitney, 1982; Whitney et al., 1983). In addition, sequence issues highlight assembly machine and tooling design problems, such as part approach directions, tolerance buildup due to prior assembly steps, access for grippers, stability of subassemblies, number of tools needed, and tool change requirements. Thus, the choice of sequence, normally considered late in the process design, really belongs in the early stages, since each can heavily affect the other. For these reasons, "determination of alternate assembly sequences" occupies the center of Figure 3.

The benefit of this approach to assembly can be seen by considering a hypothetical product of six parts. It can be built in many ways, among them bottom up, top down, or from three subassemblies of two parts each. The choice among these options depends on many factors. There are construction considerations, such as access to fasteners or lubrication points. There are assembly considerations, such as sequences whose success may be doubtful or whose failure might damage some parts. There are quality control whose failure might damage some parts. There are quality control considerations, such as the ability to test the function of the subassembly before it is buried beneath many other parts. There are process considerations, such as ability to hold the pieces accurately during critical operations. Finally, there are production strategy considerations, such as being able to make in advance some subassemblies that are common to many models, allowing final assembly to be completed quickly on the remaining parts.

In traditional industrial engineering (Taylor, 1911), a major influence on choice of assembly sequence is line balance. Relevant to manual assembly, line balance is achieved by dividing up the assembly tasks so that each worker's total task time per cycle is as close to that of the other workers as possible. To achieve line balance, the industrial engineer decides on sequences and groups the tasks and assigns the groups to the workers. It should be clear from the foregoing discussion that much broader issues can be brought to bear on sequence choices, ranging from testing options to market strategy. It should also be clear that a sequence that is good for human workers may be totally irrelevant for machines, whose strengths and weaknesses are totally different from those of people. For example, it is easy for a person to turn over a small item while passing it on to the next station. For a robot or machine to perform the turnover, it must be provided with an extra powered axis at a cost that may be considerable.

This brings us to the choice of an assembly method for each step and integration of a quality control strategy. The greatest influences on choice of method are the anticipated production volumes and the need for flexibility in model mix, part count, options, method of treating units that fail tests, and so on. There is some literature on this topic, centering mainly on the technical aspects and capabilities of different methods, but there are few detailed comparisons of accuracy or reliability. There is also a dearth of economic data. The result is that one cannot at this time make a convincing prediction of the cost and throughput of candidate assembly systems. In certain specialized cases, good estimates can be made. These cases include manual systems or those consisting of specialized "assembly machines." Even in these cases, there are few data and models, and most decision making is based on informed estimates by experienced individuals. This is a developing problem area, and much work remains to be done.

A number of computer-based tools for designing and analyzing assembly systems have been developed by the Draper Laboratory (CSDL Reports 1978–1980, R-1284, R-1406; Graves and Lamar, 1983) and by others, notably Boothroyd et al. (1982). Given adequate data, these tools permit the following issues to be addressed:

• What is the best economic mix of machines and people to assemble a given model mix of parts for a product, given each machine's or person's cost and time to do each operation, plus production rate and economic return targets?
• How much can one afford to spend on an assembly system, given an anticipated revenue stream?
• How much extra time, machines, money, or product inventory are required to meet a production rate if a certain mix of failures and repair steps can be anticipated during production?
• How can one make the trade-off between the cost of higher-quality parts and the time to unjam an assembly machine when a low-quality part gets stuck?
• What is the best way to distribute work among workstations in an assembly system?

Integrating a quality-control strategy into product and process design involves many decisions, including purchasing options and personnel policies that are beyond the scope of this paper. We will touch on only two aspects related to topics already discussed: definition of subassemblies and modular assembly-line design. We have seen that a way to define subassemblies is to define assembly stages in which an object with a definable function has been built. Since that function is related to the product's specifications, we should be able to define a test for that subassembly so that we know it will do its job when mated to the rest of the parts.

To incorporate a quality control strategy into manufacturing, it is necessary to decide which of many possible assembly stages to choose as test points. Relevant factors include how costly and how definitive the test is, whether hidden flaws could become undetectable if the test were delayed to later

stages, and how much it costs to repair or discard bad subassemblies. It is not uncommon for a tear-down at final assembly to cost as much as half the total *manufacturing* cost. A rational approach involves examining the assembly sequence to determine where each fault becomes critical and each test opportunity occurs. This study may result in a new assembly sequence.

Conclusion

Company management must establish a product strategy and encourage the design team to search for the product and manufacturing system design that best suits the product's character and meets the needs of the marketing and manufacturing approach. It should be clear that a great deal of work is required to think through the options and arrive at a good final design. Technology alone cannot create a productive manufacturing organization. Fortunately, the emergence of new system design tools, methods, and part mating models and analyses has created a new knowledge base to support this work. Furthermore, institutional changes will have to be made to allow the necessary analyses to take place. We have found that any artifact, new product, etc., can be used as a focus to rationalize the formation of the concurrency teams and to guide their activities.

RECENT DESIGN STRATEGY EXAMPLES

Automobile Factory: Success Based on Technology and Product Redesign

Volkswagen's remarkable Hall 54 was recently opened to the public. In it, Golfs and Jettas are put through final assembly with 25 percent of the steps done by robots or special machines, as compared with 5 percent in the past.

The full impact of this change can be

better appreciated by considering the conventional automotive product cycles. Many products are proposed and undergo development and prototype design, but only a few are approved. Once a product is approved, financial support for its final development is assured. At the same time, however, a product introduction date (PID) is set, usually only 24 or 36 months ahead. This date is so near that little time for rationalizing the design is available. Purchase orders for machinery must be negotiated immediately. There is little, if any, time for making a significant change in the design. There is also great reluctance to change the PID unless a major problem arises.

To make Hall 54 a success, Volkswagen (VW) obtained approval from its board of directors to delay introducing these cars for a year while "every part was examined" (Hartwich, 1985), and several significant departures from conventional automotive design practices were made. One example is the configuration of the front end. At a cost of adding one extra frame part, the front was temporarily left open so that the engine could be installed by hydraulic arms in one straight upward push. Normally a 1-minute operation or longer, requiring several men, this process is now accomplished unmanned in 26 seconds. Another example concerns the use of screws with cone-shaped tips. VW introduced these fasteners, which easily go into holes even if the sheet metal or plastic parts are misaligned, at a cost penalty of 18 percent for the screws. This innovation made robot-and-machine insertion of screws practical. In the following 2 years, use of cone-shaped screws became so prevalent in Germany that their price has dropped to that of flat-tip screws.

Automobile Factory: Success Based on Management

New United Motor Manufacturing, Inc. (NUMMI), the joint General Motors (GM)-Toyota company, has been in operation since December 1984. It builds a Toyota-designed small car (Ikebuchi, 1986). The factory has 2,500 team members, 27 Toyota managers, and 16 GM managers. The team members were carefully chosen from the group of United Automobile Workers (UAW) that were originally employed at the GM Fremont plant. The UAW, which represents the team members, agreed to many new work rules. In addition, NUMMI undertook a detailed analysis of the main causes for lower productivity and used various techniques to solve the problems that were identified in the following areas:

• Quality problems resulted from an assumption that repair is a regular process, that standardized work is not practical, that high worker absenteeism must be accepted as normal, that parts and components were regularly damaged during conveyance, and that the low quality of suppliers' parts must be tolerated.
• Low line efficiency resulted from ineffective job classifications and a centrally controlled maintenance group. Low line efficiency created an excessive inventory and a low utilization of equipment.
• Excess inventory created difficulties in inventory control, problems with engineering changes, repair problems, too many parts around line workers, and slowness of response time to problems.

These problems were solved in the following ways:

• Implementation of *jidoka*, the quality principle, in which machines or production stop under abnormal conditions, production equipment senses malfunction or substandard parts, the machine stops itself, and the line workers can stop the line.
• Employment of just-in-time (JIT) production and conveyance system that provides only the parts needed, when they are needed, in just the amount they are needed. This requires a *kanban* system and a system that reduces the die setup time.

• Achievement of high operation availability with foolproof devices, self-monitoring devices, light displays, buzzers, electrical circuit graphics boards, widely used signs, and simplified job classifications.

The basis of the NUMMI system—the involvement of the production workers—created a yardstick to ensure quality and a guide for employee training.

Radiator Factory: Success Based on Integrated Design

Nippondenso is the Delco of Japan. It builds generators, alternators, voltage regulators, radiators, antiskid brake systems, and so on. Its main customer is Toyota. Over the years, Nippondenso has learned to be Toyota's supplier, especially how to live with daily orders for thousands of items in an arbitrary model mix. To meet this challenge, Nippondenso has employed several strategies, including in-house development of manufacturing technology, jigless manufacturing methods (where possible), and the combinatorial method of meeting model mix.

The combinatorial method is the basis of the strategy, so it is described first. A product is divided into generic parts or subassemblies, and necessary variations of each are identified. The product is designed so that any combination of types of these basic parts will go together physically and functionally. If there are 6 basic parts and 3 varieties of each, then the company can build $3^6 = 729$ different models. The in-house manufacturing team cooperates with the designer of these parts so that the manufacturing system can handle each part, possibly by designing common jigging features onto them, or by advising the designers how to make the product hold itself together so that no jigs are needed.

The in-house manufacturing team provides other advantages, including protection of proprietary information concerning future product plans or design details, a reduction in overall cost, and a cohesive group that has learned the company's philosophy and knows how to contribute.

This strategy was employed in the design and manufacture of radiators (Ohta and Hanai, 1986). As shown in Figure 5, the basic parts are the core (with its basic parts, tubes, and fins), two end plates, and two plastic tanks. Cores and end plates snap together so that they do not need jigs while being oven-soldered together. The tanks are plastic and are crimped on so that prior soldering is not melted. The crimp die can be adjusted between cycles to take any tank size, so radiators can be processed in any model order in any quantity. When asked, "How much did this factory cost?" the chief engineer on this project replied, "Strictly speaking, you have to include the cost of designing the product."

WHY IS CONCURRENCY HARD TO IMPLEMENT?

Experience has revealed that so many advantages are to be gained from implement-

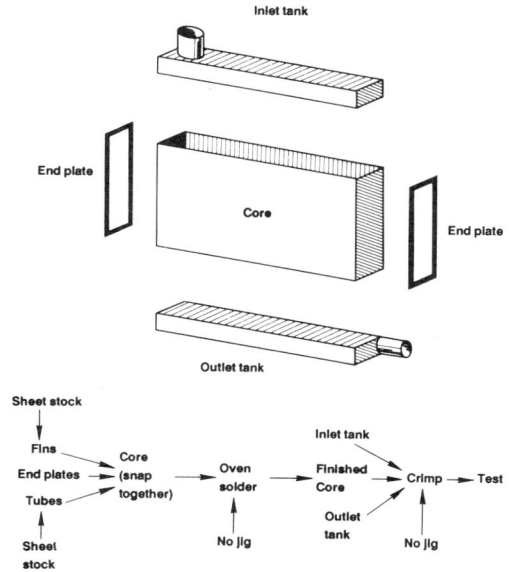

FIGURE 5 Jigless batch-size of one radiator manufacture.

ing a strong product-process link, one must ask why it is not done more often. A part of the explanation would seem to be as follows: Manufacturing is probably the most complex peacetime activity that people engage in. Together with the allied fields of finance and marketing, it can engage hundreds or even thousands of people within a single firm. A natural response to complexity is specialization, in which people acknowledge that they cannot know everything. Organizations grow up around specialization boundaries, and people must subscribe to one species or another in order to have a place. By contrast, successful implementation of the product-process link requires crossing boundaries.

Sometimes the ideas that link product and process are too new to be acceptable to established organizations. If the processes are not well understood and seem to require certain "experts" or particular conditions for their success, then conservatism against change may be a rational response.

The changes needed in people and organizations to carry out these integrated efforts can be difficult to accomplish. But, as the NUMMI example indicates, these changes must be made if the desired productivity goals are to be achieved. The National Research Council's Manufacturing Studies Board rightly devoted an entire chapter of its report *Toward a New Era in U.S. Manufacturing* to this difficult issue (National Research Council, 1986, Chapter 3).

WHAT NEW KNOWLEDGE IS NEEDED?

Broad Issues

This paper has described a new strategy for improving productivity. The strategy is based on using the assembly process as the focal point and integrator of all the complex decisions required to create a producible product. To verify, improve, extend, and implement this strategy, it is not

enough simply to try harder. A deeper understanding of the fundamental problems is needed. It will require research to identify the new knowledge that is needed as well as to fill the knowledge gaps. Although there are many ways to organize these questions, the scheme indicated in Figures 6 and 7 has been chosen. These figures deal respectively with the design of products and the design and operation of manufacturing systems. The issues are set forth in terms of increasing numbers of items (parts or workstations) and the corresponding increase in complexity.

Tools that support integrated synthesis, design, and evaluation and data bases to support these tools are needed at each level to help designers and engineers seek alternatives and to evaluate these alternatives technically and economically. As stated earlier, fabrication by traditional methods has the most advanced design methods, whereas assembly and concurrent design operate with few design tools or none at all. The lack of tools for assembly design is especially acute because the usefulness of assembly analysis in forming the total product design has been recognized only recently, and the outlines of effective strategies are only beginning to emerge.

In a larger sense, the lack of formal manufacturing design tools and computer im-

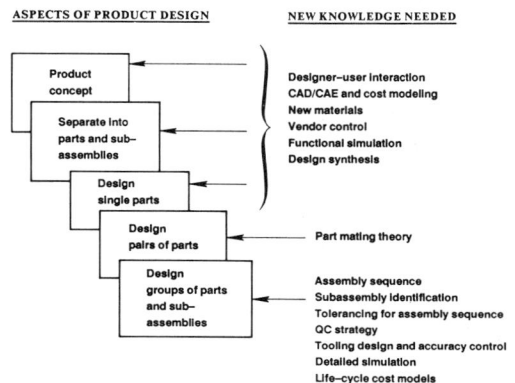

FIGURE 6 Activities and knowledge needed to support the strategic approach to product design: Part 1—Product design.

ASPECTS OF FABRICATION AND
ASSEMBLY SYSTEM DESIGN

NEW KNOWLEDGE NEEDED

Fabrication
station

Technology choice
Efficiency

Groups of fabrication
stations

Technology choice
Design, layout
Optimization
Tool and fixture management

Fabrication
systems
with transport

Process plan and sequence
Match of part types
    to technology types
Operating efficiency
Production smoothing
Scheduling
Role of people
Economic analysis

Assembly
stations

Technology choice
Design, layout, optimization
Error control

Groups of
assembly stations

Technology choice
Task assignment

Assembly system
with transport

Optimum assembly sequence
Task assignment
Tool change distribution
Scheduling for model mix
Role of people
Economic analysis

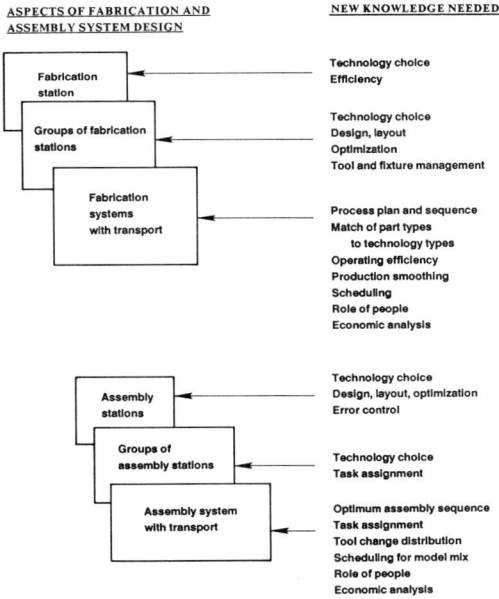

FIGURE 7 Activities and knowledge needed to support the strategic approach to product design: Part 2—Design of fabrication and assembly systems.

plementations of them is a serious deficiency for manufacturing engineers. The design engineers have a large array of computer and analytical tools at their disposal. This selection gives them a scientific as well as a psychological advantage over manufacturing engineers. The result is that product designers are better able to defend their design decisions in the face of attempts by manufacturing engineers to make the designs more producible. This result may satisfy the product designers but may result in overpriced and noncompetitive products.

On the assumption that future products will be designed by teams of product *and* manufacturing engineers, there is a need for research on the group dynamics of complex design projects. Matters of concern include sharing of data, moderating the dominance of one constituency over another, defining an effective sequence for making major decisions, and managing the process of negotiating decisions in the presence of conflicting aims.

Accompanying the design tools and methodologies must be a large array of data bases. Although many such bases are available for the design of products, more will be needed as new products are created using new materials or processes. Data bases will also be needed for describing the new processes that will be used. Finally, data bases are needed for the system design methodology itself. For example, there is a need for augmented CAD descriptions of the parts, including notations of possible assembly sequences and call-out of the relevant tolerances on jigging and gripping surfaces. There is a need for cost-tolerance data, so that the impact of various processing and assembly strategies can be assessed, and for part-mating data linked to the tolerances, so that overall assembly errors and likelihood of assembly success can be calculated. Data bases are needed for fault analysis and fault tree or other methods of representing possible product faults during assembly, so that a quality control strategy can be implemented. Finally, there is a need for data bases on the costs and capabilities of manufacturing and assembly methods and equipment that will aid in the economic analysis and synthesis of effective systems. Additional data bases, such as current warranty data and known field failure modes, may also be relevant.

Another important knowledge gap concerns performance, including performance models of both the product and the production systems, and broad aspects of performance including normal and abnormal operating modes, downtimes, repair scenarios, the supporting logistics, and so on. In many cases, the lack of adequate product performance models results in overdesigned products. In the case of manufacturing systems, performance models are often narrowly drawn with respect to local economic criteria. Too little is known about the cost-benefit relations with respect to flexibility, or how to achieve flexibility by appropriately balancing product design, manufacturing system design, use of new technology, scheduling and resource allocation, and human effort.

Also missing or immature is a strategy for educating engineering students to be effective players in such activities. Interestingly, it appears that Japanese universities are no better than American schools in this regard. Japanese companies, however, appear to compensate for this by typically giving new employees 3 years of training, with rotations throughout the company. Our suggestion is that students be convinced that a systems approach is the unifying principle, since this will best prepare them for the essential integrative nature of the activity. Once they are aware of the importance of considering many diverse factors before making a design decision, they can then concentrate on becoming expert in one particular area.

Specific Near-Term Knowledge

Progress is being made in the following specific topical areas:

• Methods of generating alternate assembly sequences for products.
• Algorithms for assessing the tolerances assigned to parts to see if they support a particular assembly sequence.
• Procedures for predicting failure modes for a product so that a test strategy can be created for each assembly sequence.
• Economic analysis methods for assessing (a) the basic assembly cost by various methods, including people and machines; (b) the cost of providing part, tool, and fixture tolerances of different accuracies; and (c) the cost of doing tests at various points in an assembly sequence, including the cost of uncovering the fault by disassembly as well as fixing it.

The status of each of these topics is discussed in the following sections. Some of the interactions are shown in Figure 8.

Generating Assembly Sequences

The importance of having a good assembly sequence has been discussed. To accomplish this requires that alternatives be generated and then evaluated in some rational way. In the past, alternate assembly sequences were frequently generated by hand, using either real parts of the product or cutouts of drawings of the parts. This cumbersome process rarely led to a large set of alternatives from which to choose. Traditionally, industrial engineers have used a diagram called a precedence diagram to represent the geometric and other constraints that limit how a product may be assembled. However, precedence diagrams do not themselves generate assembly sequences. Furthermore, many real products have constraints that cannot be represented by precedence diagrams.

On the basis of prior work (Bourjault, 1984), we have created an algorithm that will enable an engineer to generate all of the physically possible assembly sequences for a product (De Fazio and Whitney, 1987). The algorithm operates by collating the answers to a series of questions that the engineer asks regarding the assembly opportunities between related parts. The result is typically many hundreds of sequences. We also have methods for reducing this number to manageable proportions by applying judgment criteria such as how many subassemblies or flip-overs are required. As yet, we have not linked these judgments to economic criteria or to testing strategies, but it is likely that this can be accomplished.

Assessing Tolerances

In the design of a part, the accuracy of its manufacture must be specified. Some of its surfaces are important to its function, so the designer states tolerances on them for this purpose. However, tolerances must be set for additional surfaces so that assembly

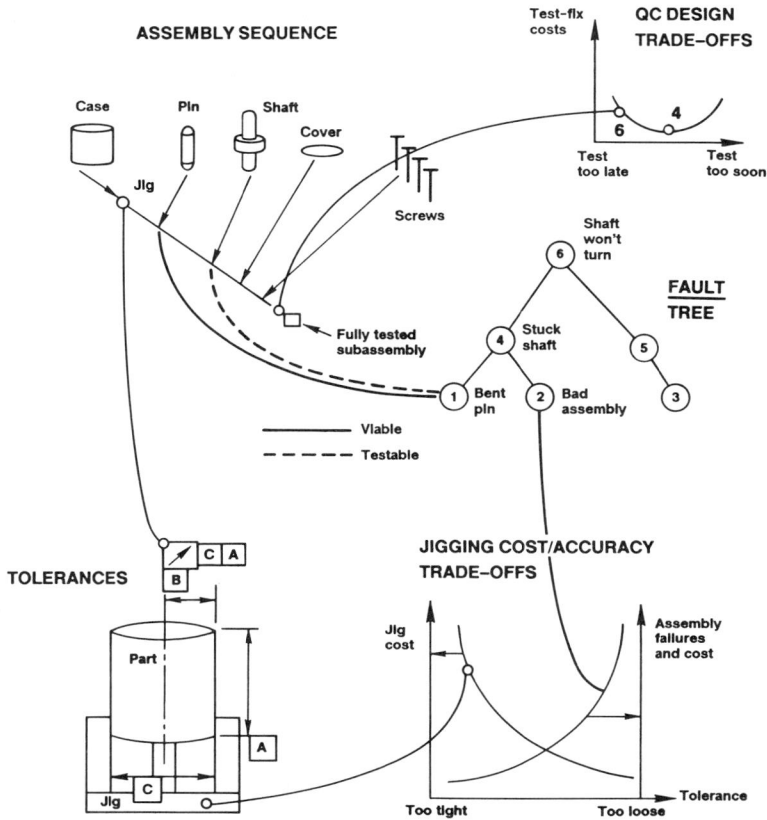

FIGURE 8   The connection between the assembly sequence and detailed part design, jig and tool design, and quality control strategy.

can take place with confidence that the parts will mate properly. There already exists a large body of theory on how far parts can be misaligned from each other and still be assembled (Whitney, 1982; Whitney et al., 1983). Thus, the requirement on the designer is to see that the surfaces on which one part rests and the other is grasped are accurately enough made and located on the parts. Naturally, depending on the assembly sequence, the resting and grasping surfaces will be different. This means that the assembly sequence must be known to the parts designer very early in the design process. By contrast, the usual practice has been to delay consideration of the assembly sequence until after the parts are designed and fabrication methods have been chosen.

Since different fabrication methods cost different amounts and are capable of making parts to different tolerances, these choices, if made without assembly process knowledge, can render an assembly sequence unrealizable.

Several kinds of knowledge and design tools are needed. First, we need better models of the costs and accuracy capabilities of different fabrication methods. Second, we need better methods and standards for establishing tolerances. Current methods can be ambiguous, so that a fabricator may not know how the part should be made or inspected. Third, many tolerances combine to create the relative locations between two parts. On a typical part, some dimensions may have turned out to be at the high end of their allowed range whereas others may be at the low end. If N dimensions or tolerances combine to create the relative locations of two parts about to be mated, then there are 2^N candidate worst cases to evaluate. Among these may be some that will prevent the parts from being assembled. Algorithms for finding these worst cases efficiently are needed.

Linking Failure Modes to Assembly Sequences

Every product can experience failures that are due to poorly made parts or problems during assembly. Some of the defects that lead to failures can be detected by inspecting the parts, whereas others do not exist until the parts are mated with other parts, fed electric power or fluid pressure, or pushed or twisted. Thus, the failures become critical at certain times during the assembly process, depending on the assembly sequence. Furthermore, although a failure may be critical at one point in the assembly sequence, it may not be testable until later. Moreover, it may again become untestable still later in the assembly. Finally, the costs of testing and of repair may depend on when the failure is detected, with earlier usually being cheaper. Thus, algorithms are needed that can relate an assembly sequence to a product's fault tree—i.e., a diagram of failures and combinations of failures as well as their manifestations at test points—so that test opportunities, test costs, and rework costs can be assigned to each candidate assembly sequence.

Economic Models of Assembly

All of the foregoing design issues have economic consequences that become part of the evaluation process. Different designs support different assembly sequences, require different tolerances on different surfaces, and require or invite various test and repair strategies. The economic choices presented must be added to those normally made when designs are evaluated and assembly methods are chosen. Although we lack the evaluation tools cited, we have already developed other tools that are useful in the design process:

• *Rework analysis:* If the test and repair points and costs of a product are known, and if the failure rates are known, the overall cost of assembly, including test and repair, can be calculated. Such calculations can include products that must be repaired twice or more and can distinguish the need for full disassembly from partial disassembly. If a different assembly sequence permits lower cost tests, or if a robot has a lower rate of product failure than a person, the cost impact of such alternatives can be calculated. The opportunity to eliminate or substantially reduce reworking has sometimes been found to save enough money to justify automation regardless of labor savings (De Fazio, 1986; Gustavson, 1986).

• *Unit cost analyses:* The cost of assembling each product unit can be calculated, given the assembly sequence, the equipment or people to be used for assembly, and the usual economic data, such as wage rates, interest rates, production quantity, and taxes (Gustavson, 1983). Conversely, given a selling price and a profit or return requirement, a maximum budget for an assembly system can be determined.

• *System syntheses:* Algorithms are available that can design assembly or fabrication systems (CSDL Reports 1978–1980, R-1284, R-1406; Graves and Lamar, 1983; Holmes, 1987). That is, given the assembly sequence and data on alternative equipment or people capable of performing each of the assembly steps, the algorithm will select people or equipment and assign the steps to them, taking into account tool costs, time to change tools, and time to transport work from station to station. The selected methods together will produce the product for minimum unit cost. In a recent study, it was shown that different assembly sequences for the same product could differ by 5 percent to 20 percent in minimum assembly cost when the algorithm had the same equipment options available (Klein, 1986).

SUMMARY

The thrust of this paper is that productive manufacturing systems can be achieved by an integrated approach to manufacturing independent of technology, and that the perceived advantages of applying advanced technology can be achieved only through this integrated approach. The integrated approach is more important than any specific technology. A new activity called the strategic approach to product design (SAPD), because of its ability to force complicated trade-offs into the open, can act as the catalyst to rationalize product and process design.

SAPD is not just a product design method but is also a way of systematizing the way people and manufacturing functions interact. It provides a basis for the creation and effective operation of concurrency teams. The method is not an end in itself. Instead, the application of the method provides insight into manufacturing systems and their interactions not currently analyzed or understood by any other method. Application of new technology without this integrated approach has proved to be disastrous (Nag, 1986).

Possible Research Initiatives

Few individuals have the necessary skills to serve effectively on the concurrency teams. In some industries the average age of those who can operate in this mode is quite high. The nation has a vested interest in capturing their knowledge, creating new knowledge, and developing methods that

will eventually allow a few individuals working at computer terminals to explore new designs with the same range of sensitivities that concurrency teams currently muster.

The assembly sequence itself is an excellent focus for the interaction between product and manufacturing engineers. The assembly sequence can act as the framework for discussions of tolerances, testing strategies, assembly methods, and various economic analyses. A number of analytical and computer tools exist to help in this process, but they are weak except in certain purely economic areas. Several new tools have been identified. A variety of data bases will be needed to support these tools. Such data bases include economic-tolerance-cost models of fabrication methods for parts, fixtures, and grippers as well as failure mode models of products to permit quality control strategies to be formulated. These new tools would allow logistic and field-support issues, as well as life-cycle costs, to be considered with respect to the total product design and manufacturing system.

Education and Institutional Impacts

SAPD, concurrency, or advanced manufacturing system technology requires engineers to have a broadly based education. This kind of education is not being offered by U.S. universities at present. Moreover, other areas with the same level of complexity, such as large-scale systems, are likewise not being supported.

Methods need to be found that allow both the present highly specialized education systems to flourish while encouraging the more broadly based training. A number of universities, with the support of the National Science Foundation and the Department of Defense, are exploring the role of centers of excellence to carry out this mission. It is not clear whether the complexity of advanced manufacturing can be adequately explored with this method.

Most of the manufacturing system design decisions needed, as well as the skills of the people who make these decisions, are currently based on experience. As industry moves from this experience base of operation to a science base, severe displacements in the ranks of middle management can be expected. Chapter 3 of the Manufacturing Studies Board report (National Research Council, 1986) explores several facets of concern and suggests that a different kind of organization will result, one more organized as a team than the conventional adversarial management-worker relationship. In such a team, all members will participate in the decisions. The NUMMI experience certainly supports this view.

ACKNOWLEDGMENT

Many of the ideas in this paper appeared in "What Progressive Companies Are Doing to Raise Productivity," by J. L. Nevins and D. E. Whitney, prepared for the Defense Manufacturing Forum, "Rethinking DoD Manufacturing Improvement Strategies," at the Institute for Defense Analyses, October 29, 1986.

REFERENCES

Akagi, S., R. Yokoyama, and K. Ito. 1984. Optimal design of semisubmersible's form based on systems analysis. ASME Paper 84-DET-87, ASME Journal on Mechanisms, Transmissions, and Automation in Design 106(4):524–530.

Boothroyd, G., C. Poli, and L. Murck. 1982. Automatic Assembly. New York: Marcel Dekker.

Bourjault, A. 1984. Contribution à Une Approche Méthodologique de l'Assemblage Automatisé: Elaboration Automatique des Séquences Opératoires. Ph.D. dissertation, Université de Franche-Comté.

Chirillo, L. D. 1982. Product Work Breakdown Structure. Maritime Administration, National Shipbuilding Research Program, U.S. Department of Transportation.

CSDL Reports. 1979–1982: R-1407, R-1537. 1974–1980: R-800, R-850, R-921, R-996, R-1111, R-1276. 1978–1980: R-1284, R-1406. Charles Stark Draper Laboratory, Cambridge, Mass.

De Fazio, T. L. 1986. Uncertainty in Unit Costs Occurring During Low-Throughput Operation of Process Systems That Include Testing and Rework. MAT Memo 1299. Charles Stark Draper Laboratory, Cambridge, Mass.

De Fazio, T. L., and D. E. Whitney. 1987. Simplified generation of all assembly sequences. IEEE Journal on Robotics and Automation. In press.

Graves, S. C., and B. W. Lamar. 1983. An integer programming procedure for assembly system design problems. Operations Research 31(3):522.

Gustavson, R. E. 1983. Choosing manufacturing systems based on unit cost. Pp. 85–104 in Proceedings of the 13th International Symposium on Industrial Robots, Chicago. Dearborn, Mich.: Society of Manufacturing Engineers.

Gustavson, R. E. 1986. A Statistical Analysis of Recycling, Rework, Yield, and Cost Reduction. MAT Memo 1300. Charles Stark Draper Laboratory: Cambridge, Mass.

Hartwich, E. H. 1985. Possibilities and trends for the application of automated handling and assembly systems in the automotive industry. Pp. 126–131 in International Congress for Metalworking and Automation, 6th EMO, Hanover, FRG.

Holmes, C. A. 1987. Equipment Selection and Task Assignment for Multiproduct Assembly System Design. S.M. thesis. Massachusetts Institute of Technology Operations Research Center, Cambridge.

Ikebuchi, K. 1986. Unpublished speech. Conference on Future Role of Automated Manufacturing, New York University.

Klein, C. J. 1986. Generation and Evaluation of Assembly Sequence Alternatives. S.M. thesis. Massachusetts Institute of Technology Mechanical Engineering Department, Cambridge.

Nag, A. 1986. Tricky technology: Auto makers discover factory of the future is headache just now. Wall Street Journal 13 May:1.

National Research Council (NRC). 1986. Toward a New Era in U.S. Manufacturing: The Need for a National Vision. Manufacturing Studies Board, Commission on Engineering and Technical Systems. Washington, D.C.: National Academy Press.

Ohta, K., and M. Hanai. 1986. Flexible automated production system for automotive radiators. Pp. 553–558 in Proceedings of the First Japan-USA Symposium on Flexible Automation, Osaka, Japan.

Olmer, L. H. 1985. U.S. manufacturing at a crossroads: Surviving and prospering in a more competitive global economy. International Trade Administration, U.S. Department of Commerce.

Opitz, H. 1967. A Classification System to Describe Workpieces. Oxford, England: Pergamon Press.

Porter, M. E. 1986. Why U.S. business is falling behind; the country is investing too little in the technology, facilities, and education it needs in today's market. Fortune 113(April 28):255–262.

Taylor, F. W. 1911. The Principles of Scientific Management. New York and London: Harper and Brothers.

Wheelwright, S. C., and R. H. Hayes. 1985. Competing through manufacturing. Harvard Business Review 63(1):99–109.

Whitney, D. E. 1982. Quasi-Static Assembly of Compliantly Supported Rigid Parts. ASME Journal of Dynamic Systems, Measurement, and Control 104:65.

Whitney, D. E., R. E. Gustavson, and M. P. Hennessey. 1983. Designing chamfers. International Journal of Robotics Research 2:3–18.

Whitney, D. E., T. De Fazio, R. Gustavson, J. Nevins, D. Selzer, and T. Stepien. 1986. Implementation Plan for Flexible Automation in U.S. Shipyards. CSDL report prepared for Todd Pacific Shipyard, Los Angeles Division, and SNAME Ship Production Committee Panel SP-10, Flexible Automation.

ADVISORY COMMITTEE FOR THE CONFERENCE

"Design and Analysis of Integrated Manufacturing Systems: Status, Issues, and Opportunities"

JAMES J. SOLBERG (*Chairman*), Director, Engineering Research Center for Intelligent Manufacturing Systems, Purdue University

W. DALE COMPTON, Senior Fellow, National Academy of Engineering

GEORGE H. KUPER, Executive Director, Manufacturing Studies Board, National Research Council

RICHARD C. MESSINGER, Vice President, Research and Development, Cincinnati Milacron, Inc.

A. ALAN B. PRITSKER, President, Pritsker and Associates, Inc.

WILLIAM P. SLICHTER, Executive Director, Research, Materials Science and Engineering Department, AT&T Bell Laboratories (Retired)

JOHN A. WHITE, Regents' Professor and Director, Material Handling Research Center, Georgia Institute of Technology

RICHARD C. WILSON, Professor Emeritus, Industrial and Operations Engineering Department, University of Michigan

CONTRIBUTORS

ERICH BLOCH is director of the National Science Foundation. Before joining the NSF, Mr. Bloch was vice president for Technical Personnel Development at IBM Corporation, which he joined in 1952. He received his education in electrical engineering at the Federal Polytechnic Institute of Zurich and the University of Buffalo. He is a member of the National Academy of Engineering.

W. DALE COMPTON became a senior fellow at the National Academy of Engineering in 1986. He joined the Ford Motor Company in 1970 and served as vice president of Research from 1973 to 1986. He received his Ph.D. in physics from the University of Illinois. He is a member of the National Academy of Engineering.

KATHY PRAGER CONRAD is a senior account manager at New England Strategies, Washington, D.C. Before taking that position she was an issues analyst in the Office of Legislative and Public Affairs at the National Science Foundation, where she developed policy recommendations for the director of the Foundation. She received a B.A. degree in biology-psychology from Wesleyan University.

THOMAS L. DE FAZIO is currently employed at the Charles Stark Draper Laboratory as a technical staff member in the Robotics and Assembly Systems Division. His major fields of interest include preliminary system and machine design, mechanics, combinatorics, and other issues of mechanical assembly, mechanics, and instrumentation. He received his B.S., M.S., and Sc.D. degrees in mechanical engineering from the Massachusetts Institute of Technology.

ULRICH FLATAU is architecture development manager, CIM Marketing Group, Digital Equipment Corporation. He joined Digital in 1974 and has worked as a senior marketing consultant for Europe and country marketing manager for the engineering market in West Germany. He received his Diplom Engineer in mechanical engineering from the University of Berlin, FRG.

F. HANK GRANT is president of FACTROL, Inc. Before forming FACTROL, he was with Pritsker & Associates. He has been actively involved in the development of special-purpose simulation languages and is currently concerned with the use of simulation tools for shop floor control and real-time scheduling. Dr. Grant received his Ph.D. in industrial engineering from Purdue University.

RICHARD E. GUSTAVSON is a technical staff member in the Robotics and Assembly Systems Division at the Charles Stark Draper Laboratory. His major fields of interest are assembly system design, manufacturing economic analyses, kinematics of linkages, and applications of trigonometry. He received his B.S. degree in mechanical engineering from the University of Michigan and his M.S. degree in mechanical engineering from Stanford University.

RICHARD W. METZINGER is a section chief in the Robotics and Assembly Systems Division at the Charles Stark Draper Laboratory. His major fields of interest are simulation, scheduling and control systems, and instrumentation. He received his B.S. and M.S. degrees in electrical engineering from the Massachusetts Institute of Technology.

ARCH W. NAYLOR is a professor of electrical engineering and computer science at the University of Michigan. He has been associated with Manufacturing Data Systems, Inc., and also with the Industrial Technology Institute in Ann Arbor, Michigan, where he worked on manufacturing software. He received his B.S. degree from the University of California, Berkeley, and his Ph.D. degree from the University of Michigan.

JAMES L. NEVINS is division leader of the Robotics and Assembly Systems Division at the Charles Stark Draper Laboratory where he directs a variety of activities involving automation. The principal focus of his work is applied research on advanced robotics, intelligent and autonomous systems, and programmable flexible automation and assembly systems. He received his B.S.E.E. degree from Northeastern University and his M.S. degree from the Massachusetts Institute of Technology.

ARTHUR J. ROCH, JR., is director of Industrial Modernization (IMOD) for the LTV Aircraft Products Group, Military Aircraft Division. The IMOD department is involved in improving productivity, lowering costs, and enhancing product quality through the use of such technology as artificial intelligence and computer-integrated manufacturing. Mr. Roch received his engineering degree from Auburn University.

JONATHAN M. ROURKE is currently employed at Arthur D. Little as a consultant in the Operations Management Section. His major fields of interest include mechanical design and manufacturing management. He received his B.S. degree in mechanical engineering from Worcester Polytechnic Institute and his M.S. degree in mechanical engineering from the Massachusetts Institute of Technology.

WILLIAM B. ROUSE is president and principal scientist of Search Technology, Inc., a company that specializes in contract R&D and consulting services in decision support and training systems for personnel in complex engineering systems. He also served as professor of industrial and systems engineering at the Georgia Institute of Technology. He received his B.S. degree in mechanical engineering from the University of Rhode Island and M.S. and Ph.D. degrees in systems engineering from the Massachusetts Institute of Technology.

LAURENCE C. SEIFERT became vice president, Engineering, Manufacturing and Production Planning at AT&T in 1987. He was director of Engineering at the Oklahoma City Works of AT&T until 1985, when he became vice president of Manufacturing Research and Development at AT&T. He holds a B.S. degree in electrical engineering from the New Jersey Institute of Technology.

DONALD S. SELTZER is a staff engineer in the Industrial Automation Division at the Charles Stark Draper Laboratory. His major fields of interest include the development of hardware and software for the control of automated systems and the design of advanced sensors. He received his B.S. degree in electrical engineering from the Massachusetts Institute of Technology.

JAMES J. SOLBERG is director of the Engineering Research Center for Intelligent Manufacturing Systems and a professor of industrial engineering at Purdue University. His research interests include stochastic processes, mathematical modeling, and manufacturing systems. He received a B.A. degree from Harvard College in mathematics and M.S. and Ph.D. degrees in industrial engineering from the University of Michigan.

RAJAN SURI is an associate professor of industrial engineering at the University of Wisconsin, Madison. His current interests are in modeling and decision support for manufacturing systems, specializing in flexible manufacturing systems. He received his B.S. degree from Cambridge University and his M.S. and Ph.D. degrees from Harvard University.

VIJAY A. TIPNIS is professor of mechanical engineering, Morris M. Bryan, Jr. Chair for Advanced Manufacturing Systems, Georgia Institute of Technology. He was the founder and president of Tipnis, Inc. His major interests are in the fields of manufacturing processes and software systems for manufacturing and process control. He received his Sc.D. degree in mechanical engineering from the Massachusetts Institute of Technology.

HERBERT B. VOELCKER holds the Charles Lake Chair in the Sibley School of Mechanical and Aerospace Engineering at Cornell University and is director of Cornell's Manufacturing Engineering and Productivity Program. He was a member of the Engineering Directorate of the National Science Foundation from 1985 to 1986. Before that he was a member of the electrical engineering faculty at the University of Rochester. He holds a B.S. degree in mechanical engineering and an M.S. degree in electrical engineering from the Massachusetts Institute of Technology and a Ph.D. degree in engineering from Imperial College of Science and Technology.

RICHARD A. VOLZ is a professor of electrical engineering and computer science and director of the Robot Systems Division of the Center for Research in Integrated Manufacturing at the University of Michigan. His interests are in distributed computer systems, manufacturing software, embedded real-time computer systems, and robotics. He received his B.S., M.S., and Ph.D. degrees in electrical engineering from Northwestern University.

JOHN A. WHITE is a Regents' Professor in the School of Industrial and Systems Engineering and past director of the Material Handling Research Center at the Georgia Institute of Technology. He received the B.S. degree in industrial engineering from the University of Arkansas, the M.S. degree in industrial engineering from Virginia Polytechnic Institute and State University, and the Ph.D. degree from Ohio State University. He is a member of the National Academy of Engineering.

DANIEL E. WHITNEY is a section chief in the Robotics and Assembly Systems Division of the Charles Stark Draper Laboratory. His research interests include robot kinematic and force control, the theory of part assembly, economic analysis of assembly systems, and the role of computers in automation. He received his Ph.D. degree in mechanical engineering from the Massachusetts Institute of Technology.

GLOSSARY

AI	artificial intelligence
AMAPS	advanced manufacturing accounting production system
ARX	account receiving executive system
ASCII	American Standard Code for Information Interchange
AS/RS	automated storage and retrieval systems
ATE	automatic test equipment
b-reps	boundary representations
CAD	computer-aided design
CAE	computer-aided engineering
CAM	computer-aided manufacturing
CAPP	computer-aided process planning
CIM	computer-integrated manufacturing
CIMA	computer-integrated manufacturing architecture
CIRP	International Institute for Production Engineering Research (orig. Collège International pour l'Etudie Scientifique des Techniques de Production Mécanique)
CNC	computer numerically controlled (machines)
C&ES-(OEM)	component and electronic systems (original equipment manufacturer)
CPU	central processing unit
CRT	cathode ray tube
CSG	constructive solid geometry
CVD	chemical vapor deposition

DDS	data dependent system
DEDS	discrete event dynamic systems
DFA	design for assembly
DFM	design for manufacturability
EBM	electron beam machining
ECG	electro-chemical grinding
ECM	electro-chemical machining
ECO	engineering change order
EDG	electric discharge grinding
EDM	electric discharge machining
EOQ	economic order quantity
ERC	engineering research center
FMA	failure mode analysis
FMC	flexible machining cell
FMC II	second-generation flexible machining cell
FMS	flexible manufacturing system
FOF	factory of the future
HIC	hybrid integrated circuit
HSM	high-speed machining
HTM	high-throughput machining
IC	integrated circuit
ICAM	integrated computer-aided manufacturing
IGES/PDES	intermediate graphic exchange standard/product data exchange specification
IM&M	information movement and management
IMPAC	integrated manufacturing planning and control
IMS	integrated manufacturing system (sometimes integrated machining system)
ISO	International Standards Organization
jidoka	the quality principle
JIT	just-in-time
kanban	"sign"—an inventory replenishment system
LTVAPG	LTV Aircraft Products Group
MAP	manufacturing automation protocol
MIS	management information systems
MMC	maximum material condition
MOVES	materials operations velocity system
MPCS-CP	manufacturing process control system—circuit packs
MPCS-EQ	manufacturing process control system—equipment
MPCS-LOT	manufacturing process control system—lot processing
MPL	machining programming/process language
MRP	material requirements planning system
MRP-II	manufacturing resource planning system
MSE	manufacturing systems engineering

| NC | numerically controlled (machines) |
| NPV | net present value |

OPT	optimized production technology
OSI	open-system-interconnect
OTC	operating telephone company

PADL	part and assembly description language
PID	product introduction date
PLC	program logic controller
ppm	parts per million
PPS	planning procurement system
PRISM	productivity improvement systems for manufacturing
PVD	powder vapor deposition
PWB	printed wiring board

| RFP | request for proposal |
| ROI | return on investment |

SADT	structural analyses and design techniques
SAPD	strategic approach to product design
SECS	semiconductor equipment communications standard
SFC	shop floor control
SPECS	synchronized production engineering control system
SQC	statistical quality control
SRD	systems requirements document

| TOP | technical and office procedures |
| TQC | total quality control (program) |

| UNICAD | unified computer-aided design |

| WIP | work in process |

INDEX

233

Jidoka (quality principle), 214
Just-in-time (JIT) production and conveyance
　　system, 48, 52, 109, 120–125, 214

K

Kanban
　　at AT&T, 25
　　in automobile industry, 48, 52, 214
　　in CIM, 31
　　feasibility analysis for, 125, 127
　　in Japan, 120, 122–123
　　models of, 126
　　versus traditional/American techniques, 120, 122–
　　　123

L

Life cycle concept
　　manufacturing systems, 123–124
　　model changes, 16
　　obsolescence, 129
　　product design, 205–207
　　product design learning curve, 203–204
　　product life, 4, 123–124
LTV Aircraft Products Group FMC, 35–45
　　cost/benefit analysis, 36–38, 43
　　equipment system and vendors, 38–44
　　operational phase and system structure, 38–40
　　planning and implementation, 35–38
　　second- and third-generation planning, 41–44
　　software for, 40–41
　　use of IMS, 35, 41–44
　　use of simulation in planning, 36

M

Machine tool industry
　　economic processes models of, 101–104
　　FMS, 110–112
　　IMS, 35, 41–44, 110–111
　　physical process control in, 96–101
Maintenance, *see* Product maintenance and repair
Management techniques and managers
　　and information needs for complex systems, 149–
　　　150
　　and innovation, 151
　　middle management, effects of advanced
　　　manufacturing, 151–152
Manufacturing Studies Board
　　report on status of U.S. manufacturing, 2, 7, 216,
　　　222
Materials costs, storage, 54
Materials handling systems
　　advances in Europe and Japan, 46
　　via automation, 56–57, 109
　　CIM, 46–59

computer-assisted planning, 109
　　development needs of, 57–58
　　human operator involvement, 53
　　in integrated manufacturing, 46–47, 52–59
　　just-in-time inventories, 52
　　lack of standardization, 53
　　material control systems, 54–55
　　research on, 46, 56
　　simulation of, 55
　　storage, *see* Inventory control
Mechanical design, *see* Product design modeling;
　　Solid modeling
Military applications
　　of FMS, 151
　　product design models, 196–197
Modeling, *see* Product design modeling; Simulations
　　and models; Solid modeling

N

National Research Council, *see* Manufacturing
　　Studies Board
National Science Foundation
　　role in manufacturing education, 221
　　role in research, 9
NC (Numerically controlled machines)
　　effects on operators, 152, 157
　　in evolution of manufacturing, 192–193, 197
　　within FMS, 38
　　linkage with CAD/CAM, 150–151
　　program verification for, 176–177
　　software systems for, 80–84
Networks
　　and CIM, 48, 65
　　layered model of CIM network, 72–78
　　manufacturing automation protocol for, 81
New United Motors Manufacturing, Inc. (NUMMI),
　　214–216

O

Operator tasks
　　effects of automation, 192
　　effects of CAD/CAM, CNC, and FMS on, 152–
　　　154, 157
　　for machine operation, 109–110
　　and task characterization and assessment, 154–158
Optimization strategies, 102–103
Organizational structure of manufacturing, 151–152
　　at AT&T, 13–14, 17–18, 20–21
　　as barriers to integration, 48–49
　　see also IM&M; SAPD

P

Packaging and containers
　　lack of standardization, 53